NEW MESSENGER RNA RESEARCH COMMUNICATIONS

NEW MESSENGER RNA RESEARCH COMMUNICATIONS

LEE B. KWANG
EDITOR

Nova Science Publishers, Inc.
New York

For permission to use material from this book please contact us:
Telephone 631-231-7269; Fax 631-231-8175
Web Site: http://www.novapublishers.com

LIBRARY OF CONGRESS CATALOGING-IN-PUBLICATION DATA

New messenger RNA research communications / Lee B. Kwang, editor.
p. ; cm.
Includes bibliographical references and index.
ISBN-13: 978-1-60021-488-2 (hardcover)
ISBN-10: 1-60021-488-6 (hardcover)
1. Messenger RNA. I. Kwang, Lee B.
[DNLM: 1. RNA, Messenger--genetics. 2 Neoplasms--genetics. QU 58.7 N532 2006]
QP623.5.M47N49 2006
572.8'8--dc22 2006037132

Published by Nova Science Publishers, Inc. ✢ New York

Contents

Preface

mRNA (messenger RNA) is the mediating template between DNA and proteins. The information from a particular gene is transferred from a strand of DNA by the construction of a complementary strand of RNA through a process known as transcription. Next three nucleotide segments of RNA, called tRNA (transfer RNA), which are attached to specific amino acids, match up with the template strand of mRNA to order the amino acids correctly. These amino acids are then bonded together to form a protein. This process, called translation occurs in the ribosome, which is composed of proteins and the third kind of RNA, rRNA (ribosomal RNA. This new and important book presents research from around the world in this dynamic field.

Chapter 1 - Advances in molecular genetics have provided a wealth of information relevant to the understanding of the mechanisms underlying both primary immunodeficiency diseases and lymphomas. This information provides the research community with insights into the different lymphocyte developmental stages. In this review we discuss expression profiling in lymphomas and primary immunodeficiencies affecting the B-cell lineage. In particular mantle cell lymphomas, chronic lymphocytic leukemias/small lymphocytic lymphomas and X-linked agammaglobulinemia (Btk deficiency) are discussed.

There is considerable hope that the microarray technology will lead to a better understanding of different disease processes. The results of a multitude of gene expression studies in malignant lymphomas clearly suggest that the vast majority of B-cell lymphomas emanate from antigen experienced cells or memory B-cells and not from naïve B-cells. These findings, together with preferential use of certain V heavy chain gene conformations, points to a role for antigen at some stages in lymphoma-development. This seems to be of particular relevance as to the discussion of cell-of-origin for low-grade lymphoma such as B-cell chronic lymphocytic leukemia, mantle cell lymphoma and marginal zone lymphoma.

The gene expression studies conducted in primary immunodeficiencies have foremost been helpful to better understand the basic biology of the disease. Microarray analysis has also been used as an exploratory tool to study gene expression in immunodeficiency diseases of unknown cause. For X-linked agammaglobulinemia, the technique has provided better understanding of the genes influenced by Btk.

Chapter 2 - Intercellular communication is a critical biological process for the orchestration of position-dependent cell differentiation, organ development and whole plant physiology. The intercellular trafficking of proteins and RNAs has emerged as a novel mechanism of cell-to-cell communication in plants. As a strategy to efficient intercellular communication by cell-to-cell macromolecule trafficking, plants evolved plant cell specific symplasmic communication networks: plasmodesmata (PD) and phloem. PD are intercellular symplasmic channels connecting neighboring cytoplasms and responsible for local molecule exchange. The phloem is the sieve-tube system that allows rapid, long-distance translocation of molecules. Taken together, PD and phloem conduits have been shown to allow the exchange of proteins and RNAs, consisting of supracellular symplasmic networking. In addition to intercellular movement of viral and viroid RNAs, numerous endogenous plant RNAs including mRNAs and siRNAs were identified to move between cells, suggesting that RNA-based signaling networks may function to control many plant processes. The isolation and characterization of mRNAs from pumpkin stem phloem provided evidences that mRNA molecules circulate throughout the plant. The functional importance of long-distance movement of mRNA through the phloem was shown in *GIBBERELLIC ACID-INSENSITIVE* (*GAI*), *CmNACP* and *MOUSE EARS* (*Me*) through grafting experiments. A tomato dominant gain-of-function mutant, *Me* was used to demonstrate for long-distance movement of its mutant mRNA. *Me* mRNAs produced in mutant stock were translocated to the wild-type scions, resulting in a change in leaf morphology. Recently, trafficking of *FT* mRNA, an important component of the elusive "florigen" signal, was suggested as an essential process in flowering controlled by day-length. The day-length signal perceived in the leaf induces *FT* expression and *FT* mRNA is translocated from the leaf to the shoot apex, where downstream floral regulatory genes are activated. These findings provide insight into the presence of a novel regulatory mechanism through cell-to-cell mRNA trafficking likely used by higher plants to integrate developmental and physiological processes on a whole-plant basis.

Chapter 3 - MicroRNAs (miRNAs) are an extensive newly recognized class of single-stranded non-coding RNAs that are evolutionarily conserved. They participate in the regulation of gene expression at the posttranscriptional level by using antisense complementarity to target mRNAs. Moreover, a combination of miRNAs and transcription factors can induce the inhibition of protein expression, a feature that cannot be achieved by conventional transcription factors alone. These ~22 nucleotide forms of miRNAs are generated by the processing of longer primary transcripts (pri-miRNAs) by the Dicer nuclease, which also processes double-stranded short interfering RNAs (siRNAs). Recent studies have shown that miRNAs have restricted expression and function. Hundreds of miRNAs have been cloned so far, and their roles have been revealed mainly in *Caenorhabditis elegans* and *Drosophila* and recently in mammals. The expression of miRNAs is under spatial and temporal regulation. Recently, brain-specific miRNAs have been found to be expressed at a limited stage of development and have specific neuronal functions. miRNAs are involved not only in the fate of neuronal cells and maintenance of tissue identity but also in the genetic etiology of neurological diseases. Elucidating the mechanisms of action of miRNAs will help not only in understanding their roles but also in developing therapies for neurological diseases in the near future. This review focuses on the roles of miRNAs in the nervous system with reference to neurological diseases.

Chapter 4 - RT-PCR assay is the most sensitive technique of mRNA determination. Two main problems of mRNA quantification arise: 1) choice of standard and 2) contamination of mRNA samples with genomic DNA. The first problem is usually resolved by using known amount of artificial RNA competitors which are co-transcribed and co-amplified together with the target transcripts. However, apart from technical difficulty of generating a competitor, the differences in amplification efficacies and sizes between the target sequence and the competitor can affect quantification of the corresponding PCR products. We have shown that genomic DNA solution can be used as external standard. The genomic DNA standards are more stable, simple in preparation and titration than artificial competitor. Influence of genomic DNA contamination is usually eliminated with intron-spanning primers. However this procedure is not applicable for quantification of intronless genes transcripts. We have worked out procedure to quantify genomic DNA contamination in cDNA samples using primers specific for gene encoding tryptophan hydroxylase-1 which does not expressed in the brain. These new techniques were successfully applied to study the regulation of the 5-HT_{1A} and 5-HT_{2A} serotonin receptor mRNA levels in brain structures of mice and rats.

Chapter 5 - Non-radioisotopic electron microscopic (EM) in situ hybridization (ISH) (EM-ISH) with biotinylated oligonucleotide probes is utilized for the ultrastructural visualization of pituitary hormone mRNA in rat pituitary cells. This EM-ISH is an important tool for clarifying the intracellular localization of mRNA and the exact site of specific hormone synthesis on the rough endoplasmic reticulum. The simultaneous visualization of mRNA and encoded protein in the same cell using preembedding EM-ISH and subsequent postembedding immunoreaction with protein A colloidal gold complex can provide an important clue for elucidating the intracellular correlation of mRNA translation and secretion of translated protein. This combined ISH and immunohistochemistry (IHC) under electron microscopy (EM-ISHandIHC) has sufficient ultrastructural resolution, and provides two-dimensional images of subcellular localization of pituitary hormone and its mRNA in a pituitary cell. The advantages of semiconductor nanocrystals (Quantum dots, Qdots) and confocal laser scanning microscopy (CLSM) enable us to obtain three-dimensional images of subcellular localization of pituitary hormone and its mRNA. Both EM-ISHandIHC and ISH and IHC using Qdots and CLSM are useful for understanding the relationship between protein and mRNA simultaneously in two or three dimensions.

Chapter 6 - Currently available tumor markers for hepatocellular carcinoma (HCC) are α-fetoprotein (AFP), lens culinaris agglutinin-reactive AFP (AFP-L3), and des-γ-carboxy prothrombin (DCP). However, their diagnostic potential cannot surpass abdominal ultrasonography (US) as modalities to detect small HCC at an early stage, resulting in a potential delay in its diagnosis. There is a need to develop an alternative sensitive marker to improve the early detection of HCC. We here introduce a newly developed quantitative detection method for serum hTERT mRNA, which has clinical significance in HCC diagnosis.

In 64 patients with HCC, 20 with liver cirrhosis, 20 with chronic hepatitis, and 50 healthy individuals, we measured serum hTERT mRNA by using the newly developed real-time quantitative RT-PCR with SYBR Green I. Briefly, we examined its sensitivity and specificity in HCC diagnosis, clinical significance in comparison with other tumor markers, and its correlations with the clinical parameters by using multivariate analyses and the Friedman test.

Serum hTERT mRNA levels were higher in patients with HCC than those with chronic liver diseases. We demonstrated that hTERT mRNA expression was independently correlated with clinical parameters such as tumor size, number and degree of differentiation ($P<0.001$ in each). The sensitivity/specificity of hTERT mRNA in HCC diagnosis showed a 88.2%/70.0% ratio. hTERT mRNA proved, as expected, to be superior to AFP mRNA , AFP and DCP in HCC diagnosis. More importantly, hTERT mRNA in serum was correlated with that in HCC tissue.

We report in this review that serum hTERT mRNA is a novel and available marker for HCC detection. We are also carrying out a large-scale study regarding HCC diagnosis. Furthermore, this method will be applicable for a number of hTERT-positive malignancies, and for the quantification of the expression of RNA genes or genes with a weak protein expression level.

Chapter 7 - Pyrimidine analogue 5-fluorouracil (5-FU) is a standard chemotherapeutic agent in treating colorectal cancer (CRC). Occasionally fluorodeoxyuridine (FUdR) has been used; however, the third analogue fluoro uridine (FUR) has never been used in the clinic. In this study we compared the anti-tumorigenic activity of the above agents *in vitro* and *in vivo*. The cultured cells were protected from the cytotoxic effects of the FUdR by deoxy thymidine and FUR by uridine. Cytotoxicity of these agents as analyzed *in vitro* by cell culture demonstrated FUR to be the most potent anti-tumor agent. *In vivo* we tested the effect of 5-FU and FUR on tumor bearing mice. Intra-peritoneal administration of 5-FU or FUR proved that FUR is fifteen times more effective than 5-FU. Severe inhibitory effect of FUR on TS expression and its RNA mediated cell kill may explain its superiority over the other two fluoropyrimidines tested *in vivo*.

Chapter 8 - Skeletal muscle is the biggest organ in the human body and its role is to move the body. Muscle growth and strengthening is important theme on sports or clinical medicine. Mechanical stretch has been considered to be a growth stimulation for muscle cells. Continuous stretching increased muscle weight in vivo and repetitive stretching were also reported to induce cellular growth for cultured cardiac muscle cells. We have investigated the effect of repetitive passive stretch to muscle tissues using molecular biological methods. Passive repetitive stretching induce the expression of immediate early gene, myogenic transcription factor to growth the skeletal muscles. Recent studies clarify the mechano-transduction and mechanisms about muscle growth with gene expressions. Mechanical stretching to skeletal muscle stimulates the mechanoreceptors on the surface of muscle cells and activates the intracellular signal pathway to gene activation. Then, gene expression of immediate early gene, myogenic transcription factor, growth factor, muscle structure gene were occurred and muscle growth were obtained. We introduce mechanical stimulation induce the muscle growth with expressions of mRNAs.

Chapter 9 - The epidermal growth factor receptor (EGFR) is frequently amplified and mutated in human cancers and has been identified as a promising target for new anticancer strategies. Overexpression of the EGFR by tumor cells is associated with increased cellular proliferation, motility, adhesion, tumorigenicity and invasive capacity, induced angiogenesis as well as inhibition of apoptosis. Targeting strategies directed against wild-type or mutant EGFR include small molecule tyrosine kinase inhibitors, monoclonal antibodies, immunotoxin conjugates and various classes of antisense oligonucleotides. Among several mutations of the *EGFR*, a specific 801-base pair deletion mutation (termed variant III [vIII] or Δ*EGFR*) involving the extracellular receptor domain commonly occurs in various malignant neoplasms but not in normal tissues and thus represents an outstanding molecular target. The truncated receptor is constitutively (ligand-independently) activated and defective in downregulation due to low rates of receptor endocytosis. This chapter focuses on the promise and potential limitations of specific ribozyme-mediated inhibition of Δ*EGFR* messenger (m)RNA expression as a gene therapeutic approach to ΔEGFR-expressing human malignancies.

Chapter 10 - The internal ribosome entry site (IRES) is a part of the mRNA sequence which is able to attract the eukaryotic ribosomal initiation complex and to directly promote the initiation of protein synthesis independently of the presence of 5'-terminal 7mG cap. RNA structures bearing the IRES activity were first discovered in certain eukaryotic viruses where they very often play a pivotal role in viral strategies, allowing the viral invader to overcome the overall decrease of the host protein synthesis caused either by viral proteins or by the cellular antiviral defense system. Although the IRES segments and thus the cap-independent translation initiation were first described in viruses, extensive evidence has appeared in the past few years that a similar principle of the translation initiation is utilized also by some cellular mRNAs. Demonstration of IRES activity of a particular RNA region is not a simple task. A proper design of the experiment and a careful selection of the controls – excluding artificial signals generated by leaky scanning, ribosome hopping and undesirable cryptic transcription, splicing or physical breakage at the hot-spots – are very important. A number of false positives described in the literature as well as difficulties in designing appropriate controls have become the major stimuli for creating IRESite – the publicly available manually annotated database of experimentally verified IRES structures. This chapter presents the current status of the IRESite database (http://www.iresite.org), the complete list of known viral and cellular IRESs as well as novel results obtained from the comparative analyses of IRES segments accumulated to date. The article also presents a brief description and comparison of other available databases containing IRES and 5' untranslated region (5'-UTR) related information.

Chapter 11 - Programmed ribosomal frameshifting is a common way of achieving the production of different proteins from the same mRNA, particularly in retroviruses. The efficiency of the RNA-encoded signals that induce ribosomes to switch reading frame, is usually expressed as the frameshift efficiency, *i.e.*, the percentage of the ribosomes that, upon reaching the frameshift signal, continue in the altered reading frame. In practice, this is determined either *in vitro* or *in vivo* as the amount of shifted product, divided by the sum of the shifted and non-shifted products. In the course of a study on the properties of the frameshift signal from *Simian retrovirus 1* RNA, the frameshift efficiency thus determined was found to change dramatically with time in an *in vitro* time-course experiment. Here we show that these variations in apparent frameshift efficiency are not due to varying behavior of the ribosomes at the frameshift signal, but to factors other than frameshifting, affecting the relative amounts of the two protein products. A kinetic model is presented, incorporating besides a constant probability of frameshifting also the kinetics of translation elongation over the two reading frames and the stability of the two protein products. Using our experimental data, we demonstrate that fitting the results from a time-course experiment to this model allows one to compensate for these two factors and thus arrive at more reliable values for the frameshift efficiency without the need for additional, dedicated experiments.

Chapter 12 - Monocytes and monocyte-derived macrophages are hard-to-transfect cells. Transfection of the human monocytic cell lines U937 and THP-1 with DNA encoding enhanced green fluorescent protein (eGFP) using standard transfection methods does not yield detectable levels of eGFP positive cells. However, nucleofector technology as a novel nonviral, electroporation-based method results in 30-50% DNA-transfected U937 or THP-1 cells, suggesting that this method is superior to others. Transfection efficiencies up to 80% without significant cell toxicity can be obtained when U937 and THP-1 cells are nucleofected with capped and polyadenylated mRNA specific for eGFP. In murine J774A.1 macrophages, introduction of DNA causes massive apoptosis without eGFP expression, whereas nucleofection with mRNA yields up to 75% eGFP positive cells without significant induction of cell death. Analogous with J774A.1 cells, primary peritoneal mouse macrophages can be easily transfected with mRNA (up to 60% eGFP positive cells) but not with DNA. Nucleofected cells start to express eGFP after an incubation period of only 2-6 hours. Despite successful transfection-driven eGFP gene expression, monocytes and macrophages fail to express for unknown reasons many other gene products upon mRNA transfection. In conclusion, transfection of monocytes and macrophages with eGFP mRNA is superior to DNA gene transfer, most likely because mRNA is less toxic and nonapoptotic as compared to plasmid DNA.

Chapter 13 - It is widely suggested that a eukaryotic mRNA typically contains one translation start site and encodes a single functional protein product. However, according to the linear scanning model, 40S ribosomal subunits can either initiate translation at the 5'-proximal AUG codon in a suboptimal context or miss it and initiate translation at downstream AUG(s), thereby producing several proteins. Earlier experiments suggested that some of these additional protein variants could be functionally important. I analyzed the potential alternative translation start sites in a sample of mouse cDNAs. It was found that 42% of the sample was characterized by AUG-containing 5'-untranslated regions and 14% were characterized by a suboptimal context of an annotated start codon. It is likely that many mouse mRNAs can additionally encode protein products either unrelated to annotated proteins or representing N-end truncated or N-end extended variants of the annotated protein. Statistical analysis of alternative ORFs features also demonstrated their potential functionality. In particular, there is a considerable difference in predicted subcellular localizations of many annotated mouse proteins and their potential N-end extended or N-end truncated variants. These results suggest that in mammalian cells additional AUG codons can frequently be used as alternative translation start sites to synthesize proteins possessing new functional properties. The correctness of translation start site(s) annotation, possible polypeptide heterogeneity, and the related mRNA features are discussed.

Expert Commentary

Marc-Eric Halatsch

Regardless of which mRNA antisense approach is applied to treat human cancers, the challenge of ensuring specificity of targeting remains paramount because of the potentially detrimental effects associated with nontargeted alteration of gene expression. In this regard, the 801-base pair deletion-mutant variant of the epidermal growth factor receptor (termed variant III [vIII] or ΔEGFR) mRNA constitutes an ideal target, because the fusion junction is unique to the aberrant receptor and absent in normal tissues. While short interfering (si)RNAs are generally more potent than ribozymes in downregulating the expression of target genes, their application is more prone to off-target effects, which may be a matter of serious concern. Against this background, the high specificity of ribozymes and the great effectiveness of siRNA render these two technologies complementary rather than competing.

In: New Messenger RNA Research Communications
Editor: Lee B. Kwang, pp. 1-23 ISBN 978-1-60021-488-2

Chapter I

Gene Expression in Lymphomas and Immunodeficiencies Provides New Insights into B-Lymphocyte Development

***Jessica M. Lindvall*[1,3], *K. Emelie M. Blomberg*[1], *Birger Christensson*[2], *Birgitta Sander*[2,1] *and C. I. Edvard Smith*[1,*]**

[1] Clinical Research Center, [2] Department of Laboratory Medicine, Division of Pathology, Karolinska Institutet, Karolinska University Hospital Huddinge, SE 141 86 Stockholm, Sweden

[3] Current address: Department of Informatics, University of Oslo, PO Box 1080 Blindern, NO-0316 Oslo, Norway

Abstract

Advances in molecular genetics have provided a wealth of information relevant to the understanding of the mechanisms underlying both primary immunodeficiency diseases and lymphomas. This information provides the research community with insights into the different lymphocyte developmental stages. In this review we discuss expression profiling in lymphomas and primary immunodeficiencies affecting the B-cell lineage. In particular mantle cell lymphomas, chronic lymphocytic leukemias/small lymphocytic lymphomas and X-linked agammaglobulinemia (Btk deficiency) are discussed.

There is considerable hope that the microarray technology will lead to a better understanding of different disease processes. The results of a multitude of gene expression studies in malignant lymphomas clearly suggest that the vast majority of B-cell lymphomas emanate from antigen experienced cells or memory B-cells and not from

[1] Corresponding authors, e-mail address: *edvard.smith@crc.ki.se* Telephone: +46-8-58583651. Fax: +46-8-58583650 and *birgitta.sander@ki.se* Telephone: +46-8-58587843. Fax: +46-8-58581005.

naïve B-cells. These findings, together with preferential use of certain V heavy chain gene conformations, points to a role for antigen at some stages in lymphoma-development. This seems to be of particular relevance as to the discussion of cell-of-origin for low-grade lymphoma such as B-cell chronic lymphocytic leukemia, mantle cell lymphoma and marginal zone lymphoma.

The gene expression studies conducted in primary immunodeficiencies have foremost been helpful to better understand the basic biology of the disease. Microarray analysis has also been used as an exploratory tool to study gene expression in immunodeficiency diseases of unknown cause. For X-linked agammaglobulinemia, the technique has provided better understanding of the genes influenced by Btk.

Keywords: B-lymphocyte, B-cell development, gene expression profiling, Primary immunodeficiency (PID), X-linked agammaglobulinemia (XLA), Bruton tyrosine kinase (Btk), Mantle cell lymphoma (MCL), B-cell chronic lymphocytic leukemia (B-CLL), Cyclin D1 mRNA splice variants.

INTRODUCTION

Improved definition of the B-lymphocyte developmental pathway is profoundly aided by the analysis of lymphomas and antibody deficiencies. The human genome consists of 20,000-25,000 protein-coding genes [1] and translation, post-translational modifications allow for even greater diversity. In addition there are non-protein-coding genes, many of which influence developmental process. Also, in a given cell not all genes are expressed simultaneously. As an example, using transcriptional profiling, the total number of different transcripts in a B-lymphocyte is in the range of 5,000 [2]. Many of these genes are house-keeping genes, but even among this population of genes there can be characteristic, developmental differences [3]. Important steps in B-cell development that are vulnerable and susceptible to dysregulations causing immunodeficiencies and malignant lymphoma have been identified using gene expression profiling. In tumors the accumulation of transformed cells highlights populations of cells for which the normal counter parts are rare and therefore difficult to study. Although useful, malignant cells are also characterized by both gain- and loss-of-function genetic alterations, which may result in a phenotype distinguishing the lymphoma cells from their non-malignant counterparts. A diverse array of genetic events from the stem cell and onward has now been elucidated in B lineage malignancies. Gene expression profiling of lymphomas, has provided evidence for the existence of tumor types not previously recognized and the genetic signatures indicate molecular pathways of importance for pathogenesis of both small B-cell lymphomas and aggressive lymphomas [4, 5].

Conversely, in PIDs different inherited mutations are manifested as characteristic blocks at distinct developmental stages resulting in specific phenotypes. As for the primary immunodeficiency area, the application of gene expression profiling has allowed a better understanding of these disorders at a molecular level [2, 6-9]. There is also an example where the affected gene was first identified through microarray analysis, namely the CD3δ-chain deficiency [10]. Thus, expression profiling may be used to hunt for the primary cause of a PID.

In the following sections we will summarize recent findings in studies of lymphomas and PIDs with emphasis of the low-grade B-cell lymphomas, B-cell chronic lymphocytic leukemia (B-CLL)/small lymphocytic lymphoma (SLL), mantle cell lymphoma (MCL) and marginal zone lymphoma (MZ) and of the antibody deficiency disease, X-linked agammaglobulinemia (XLA).

We will also discuss how recent insights into B-lymphocyte development may increase the understanding of lymphoma development and antibody deficiencies. Results obtained using gene expression profiling are brought to light and summarized below. In figure 1 we have outlined B-lymphocyte development with special emphasis on the stages where these diseases are manifested.

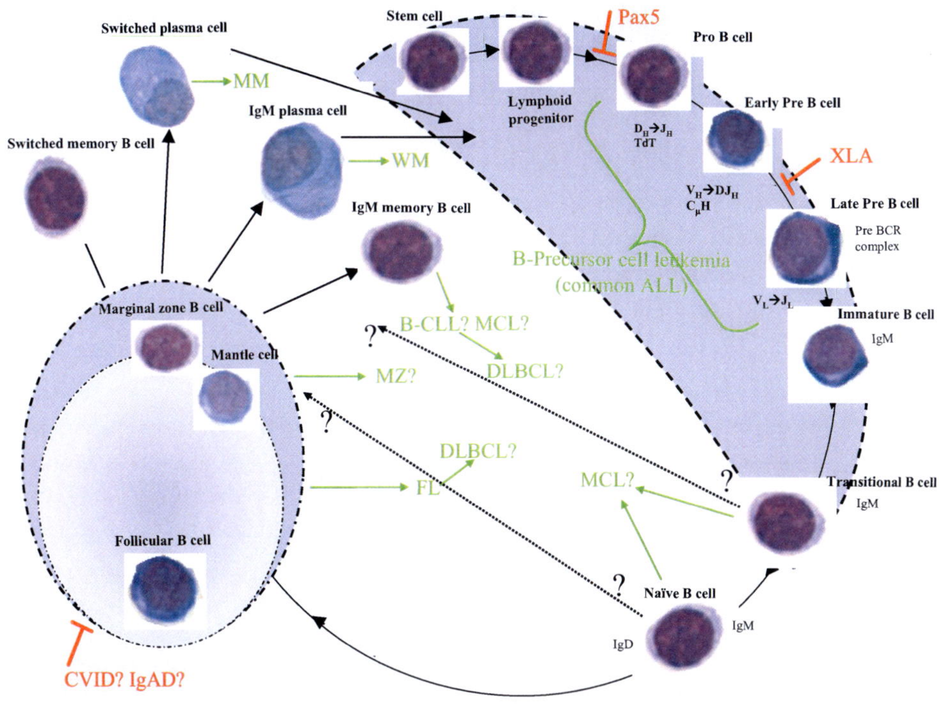

Figure 1. B-cell development and differentiation with related immunodeficiencies and B-cell lymphomas. B-cells develop from lymphoid progenitors to transitional B-cells in the bone marrow. A key regulator of B-cell development is Pax5, which when absent will cause a block after the early pro B-cell stage (shown in red). B-cells leave the bone marrow as transitional B-cells and travel into the peripheral lymphoid organs. Naïve B-cells can be recruited either to the B-cell follicle or differentiate through extra follicular pathways. Both switched plasma cells, derived from the follicular pathway, and IgM plasma cells will return to the bone marrow. The different immunodeficiencies are presented in red and the B-cell lymphomas are shown in green. MCL; mantle cell lymphoma, FL; follicular lymphoma, MZ; marginal zone lymphoma, MM; multiple myeloma, WM; Waldenstroms macroglobulinemia, DLBCL; diffuse large B cell lymphoma.

B-LYMPHOCYTE DEVELOPMENT AND DIFFERENTIATION

Haematopoietic stem cells are thought to progress through an irreversible cascade of differentiation steps as they commit to a particular cell lineage. This is a very complex process that depends on a number of B-cell intrinsic factors that need to be expressed as well as extrinsic components found in the local environment at the time of commitment. B- and T-lymphocytes provide long-term immunological memory and highly specific effector functions in the immune defence to foreign antigens. For that purpose they carry antigen-specific receptors, the ligation of which regulates diverse responses such as proliferation, differentiation and cell death depending on their developmental stage and the microenviromental context. For B-cells, there are two major phases of development; an initial phase that is antigen independent, in which a diverse repertoire of antigen specific B-cells develops in the bone marrow. The second phase occurs when B-cells are stimulated by antigen to undergo clonal expansion in the peripheral lymphoid system. During B-lymphocyte development in the bone marrow, each B-cell randomly rearranges its Ig gene segments to form one complete, productive heavy-chain gene and one complete, productive light-chain gene. This is to ensure that all B-cell antigen receptors (BCRs) on a single B-cell have the same antigen-binding site, whereas the total population of generated B-lymphocytes has a diverse repertoire of antigen-specific antibodies.

During the second phase of B-lymphocyte development, taking place in the peripheral lymphoid organs, the immature B-cell that recognizes self-antigen initiates strong BCR signaling resulting in receptor editing or apoptosis. The next step is for the transitional B-cells (T1 and T2) to differentiate into mature circulating B-cells. These transitional B-cells must then pass through multiple checkpoints before entering the mature B-cell pool. Thus, before becoming mature B-lymphocytes, the transitional B-cells are either selected to the follicle or the mantle zone. The follicular B lineage cells recirculate continuously. At the same time they are being enriched for specificities common to T-dependent antigens. Mantle zone B-cells, on the other hand, are static and are enriched for specificities common to T-independent antigens. The mature B-lymphocytes circulate throughout the body and encounter foreign antigens.

In addition to activating B-cell immune effector functions, signaling through the BCR also regulates B-cell development, proliferation, survival and elimination of auto-reactive B-cells. Signal transduction through the BCR is an absolute requirement for the selection and development of B-lymphocytes at multiple checkpoints. How can the BCR on one hand provide survival and differentiation and on the other hand induce cell death? The outcome depends on the maturation state of the B-cell, the level and duration of the BCR signal and also on signals from other receptors such as CD19 and CD40.

The initiation of BCR signaling follows upon antigen cross-linking causing the cytoplasmic domains of Igα and Igβ to become phosphorylated on their respective immunoreceptor tyrosine-based motifs (ITAMs). When phosphorylated, these motifs recruit and activate Src family tyrosine kinases. This alone is sufficient to commence the BCR signaling. One of the prevailing models is that BCRs exist as monomers on the surface of B-lymphocytes and that antigen binding to the receptor causes the BCRs to cluster allowing the Src family kinases to phosphorylate tyrosine residues on nearby BCRs and by this create binding sites for other kinases [11]. Once the BCR is activated, a cascade of proteins becomes phosphorylated, activating downstream components of the BCR signaling pathway. In contrast to this "clustering" view, an alternative model is that many receptors exist as pre-formed multimeric complexes. Binding of the ligand forces the complex to reorganise and by these means allow initiation of signaling downstream. This model has also been suggested for e.g. the tumor necrosis factor receptors and the erythropoietin receptor [12, 13]. Yet another model suggests that BCRs, before antigen binding, also exist as oligomeric complexes [14]. Here, within the same oligomer, BCR associated Src kinases continually phosphorylate ITAMs on other BCRs. This allows the recruitment of other cytoplasmic kinases to the complex and further activation of downstream signaling. The major signaling pathways activated by BCR are controlled by PI3K, PLC-γ2 and the Ras, Rac GTPases (reviewed in [15]). A key feature of all of these signaling pathways is the recruitment of cytosolic signaling proteins to the plasma membrane. These interactions among molecules in the plasma membrane, promote the assembly of different complexes resulting in the activation of downstream targets (reviewed in e.g. [16]).

The Microarray Technology for Detection of Gene Expression at the mRNA Level

In contrast to the traditional molecular technologies, where only a limited number of genes are analyzed in one assay, the microarray method allows a global view of gene expression. The overall gene expression profile provides a snapshot of the average expression of genes in the cell populations contained in the analysis. The microarray technology has been widely used in studies of lymphomas and solid tumors, drug response studies as well as basic research in immunology [4, 7, 17-19].

Currently, one of the most used gene expression profiling methods is the Affymetrix™ technique where arrays of probes are synthesized directly on glass chips using photolithography. However, nowadays there is an "array of arrays" on the market, all widely used by the scientific community. For instance, one of the most commonly used arrays is the cDNA array, which uses full-length cDNA clones spotted either on filters or on specially coated glass slides. This approach requires simultaneous analysis of two biological samples, a test and a control, each labelled with a different fluorescent dye, Cy3 (green) and Cy5 (red). Of particular interest is the Lymphochip, a gene chip specially designed for the analysis of lymphoid cells. Many studies using the Lymphochip focused on finding the molecular mechanisms causing human lymphoid malignancies.

Current densities of commercially available Affymetrix™ arrays allow measurements of up to 40,000 probe-sets or transcripts on one single array for a particular species. These arrays contain both known genes and so-called expressed sequence tags (ESTs). Almost all the publicly available EST sequences can be monitored in either human or mouse samples relatively easily by *in silico* analysis.

ONCOGENIC EVENTS IN B-CLL AND MCL

Small B-cell lymphomas comprise approximately one-third of non-Hodgkin lymphomas of B-cell type. This group includes such entities as B-cell chronic lymphocytic leukaemia/small lymphocytic lymphoma (B-CLL/SLL), mantle cell lymphoma (MCL), marginal zone lymphoma (MZL) and follicular lymphoma (FL). It is important to distinguish these lymphoma entities since their clinical characteristics with respect to therapy and survival are different. Based mainly on morphology and growth pattern, these specific lymphoma entities have been named after their proposed normal B-cell counterparts, indicating possible cells-of-origins. Thus, FL in many instances has a follicular growth pattern that imitates reactive B-cell follicles.

However, in other types of small cell lymphomas such as B-CLL/SLL and MCL the normal B-cell counterpart has not yet been well defined. It is well established that B-cells in the germinal centre of the B-cell follicle are induced to proliferate in response to stimulatory signals from antigen and co-stimulatory T-cell help. During this follicular response, affinity maturation and Ig heavy chain class switching also occurs. Following introductions of mutations in the antigen combining regions of the B-cell receptor genes, B-cells ending up with increased antigen binding capacity are positively selected. The presence or absence of somatic mutations in immunoglobulin genes has thus been considered indicative of whether a B-cell has passed through germinal centre differentiation or not. The mutational status of the immunoglobulin genes in malignant lymphoma has therefore been used, together with the expression of activation markers, to indicate whether a particular lymphoma entity has emanated from "pre-germinal", possibly naïve B-cells or from "post-germinal" B-lymphocytes. Since MCL in the majority of cases has non-mutated IgVH genes it was hypothesized that this lymphoma stems from naïve B-cells. In B-CLL/SLL on the other hand approximately 60% of the tumors have evidence for somatic mutation in the immunoglobulin genes and could therefore possibly stem from post-germinal B-cells. Recent work, however, has suggested that in some instances somatic mutation might also occur outside germinal centers, such as in the splenic red pulp [20, 21]. Therefore, it might be informative to use other tools, such as gene expression profiling to compare the various normal B-cell populations with the lymphoma entities in order to clarify possible cells of origin. The rationale for this approach is that normal B-cell populations exhibit specific gene expression signatures that in part are preserved in the malignant lymphomas. In addition, oncogenic events such as chromosomal translocations, chromosomal- or gene- amplifications and deletions affect proliferation, cell survival and differentiation and can possibly also be identified by specific gene expression profiles.

MCL is characterized by a chromosomal translocation where the cyclin D1 gene on chromosome 11 becomes translocated to the immunoglobulin heavy chain locus on chromosome 14, the t(11;14)(q13;q32). This translocation results in increased expression of cyclin D1 protein and disturbed cell cycle regulation. Detailed studies of the translocation breakpoints can provide information concerning at what time-point in B-cell differentiation the breaks occur. Such studies have suggested that the t(11;14)(q13;q32) occurs mainly during the Ig gene rearrangement process in B-precursor cells in the bone marrow, during which one D- and one J segment of the Ig locus are joined [22, 23]. The translocation breakpoint in MCL would therefore suggest that the t(11:14) translocation is an early event in MCL pathogenesis which takes place at the pre-B-cell stage and is followed by other oncogenic and selective events (see below) later on during tumorigenesis.

In B-CLL there is not a specific gene translocation. Instead, in this lymphoma there are certain patterns of genomic aberrations such as chromosome gains and deletions that provide prognostic information, such as deletion of chromosome 11q14 or 17p13 and trisomy 12 that are associated with short survival time [24-26]. Gene expression profiling of these subgroups in B-CLL have shown different gene expression profiles that only in part can be explained by a gene dosage effect provided by genes located in the aberrant chromosomal regions [27]. Apart from translocations and numeric changes there are additional genetic changes that in part are common between B-CLL and MCL. Both in MCL and CLL mutations of the ataxia telangiectasia gene (ATM) is frequently seen as well as TP53 aberrations. However, the molecular mechanisms causing B-CLL and MCL are largely unknown as is the progenitor cell(s) for the two diseases.

It has long been assumed that B-CLL and MCL are derived from human B-cell subsets constituting counterparts of the mouse B-1 subset. Moreover, since both MCL and B-CLL express the CD5 molecule it has been suggested that they arise from a B-cell population similar to the B-1a cells in the mouse. In mice at least four subpopulations of mature B-cells have been defined namely, recirculating follicular B-cells (B-2 cells), marginal zone B-cells (MZ), enriched in the marginal zones surrounding the lymphoid follicles and B-1 lymphoid cells that are divided in B-1a cells expressing CD5 and B-1 cells that are CD5 negative (reviewed in [28, 29]). The mouse B-1 cells produce polyreactive antibodies preferentially recognising self-antigens and human B-cells suggestively corresponding to B-1 cells are found in increased levels in some autoimmune diseases such as rheumatoid arthritis. The repertoire of the B1-cells is restricted and there is limited, or no, somatic hypermutation of their immunoglobulin genes. Often microbial antigens are recognized, which suggests that these cells provide the natural antibodies. Since B-CLL and MCL express CD5 and since a substantial fraction of these lymphomas have non-mutated immunoglobulin variable heavy chain genes (IgVH) it has been hypothesised that the normal counterpart for these lymphomas is a human B-lymphocyte similar to mouse B-1a cells.

Experiments, using transgenic mice expressing self-antigens, points to a role for surface immunoglobulin signaling, either during the precursor stage or later during differentiation or expansion, for generating B-1 cells [30, 31]. Importantly, the development of B-1a cells is dependent on the specificity of the antigen receptor. Thus, it seems that if a precursor B-cell expresses immunoglobulin receptors which bind self-antigens or T-independent antigens, this precursor B-cell could mature into a B-1a cell (reviewed in [28, 29]). This B-1a cell does not pass germinal centres or exhibit a high degree of somatic hypermutation in the variable region of the heavy chain of the immunglobulin gene (VH gene).

Like B-1a cells, approximately 40% of B-CLL and the majority of MCL have do not show hypermutation in the variable region ((non-mutated VH genes) and could thus be considered "antigen-naïve". However, there are several lines of evidence that point to a role of antigen also in CLL and MCL with non-mutated VH genes. During VDJ recombination one out of approximately 52 functional VH family genes is stochastically used to join the previously formed DJ complex. In both mutated and non-mutated B-CLL and MCL certain

VH genes are reported to be "over-represented" as compared to normal peripheral blood B-cells. However, there is very limited knowledge as to actual frequencies of VH gene usage in normal individuals at different ages. One can thus not with any higher degree of certainty state that some VH genes are "over-represented". Still, there is undoubtedly a marked accumulation of certain VH genes such as VH1-69 that has been associated with non-mutated B-CLL [32-34]. Similarly, VH3-21 was detected among mutated B-CLL cases [35]. Intriguingly, detailed analysis of the B-cell receptor genes reveals that in a substantial number of cases with a particular VH gene usage also the hypervariable regions are highly similar between cases [36-38]. Moreover, also a restricted immunoglobulin light chain usage has been found. Taken together, these findings strongly point towards a role for antigen in the development of CLL and MCL.

In humans the IgM+, IgD+, CD27+ B-cells in blood are responsible for T-cell independent immune reactions, such as rapid responses to encapsulated bacteria [39]. In contrast CD27- B-cells that have non-mutated immunoglobulin genes have been considered antigen naive. Recently, it was suggested that the blood IgM+, IgD+, CD27+ cells correspond to circulating splenic marginal zone B-cells that are rapidly activated during T independent immune responses [39]. Intriguingly, this population, that was detected in small children and has mutated immunoglobulin genes already at one year of age, was selected and underwent further hypermutation during immunization with vaccines against encapsulated bacteria [39]. Gene expression analysis showed that this cell-population, in comparison with B-cells lacking IgD and/or CD27, expressed CD1d and CD1c that are involved in presenting lipid antigens.

It thus appears that memory B-cells of IgM type carrying limited somatic mutations in immunoglobulin genes may be generated by a, possibly T cell independent extra follicular pathway.

Gene Expression Profiling in B-Cll, MCL and MZL

Gene expression profiling has recently been used to compare the transcriptomes in mouse B-1a cells and conventional B-2 cells. Among genes differentially expressed in B-1a cells compared to splenic B-2 cells were CD72, and CD1 [40]. By gene expression profiling, CD72 has been found to be highly expressed in human mantle zones [41] but down regulated in MCL compared to normal B-cells [42]. CD1a was reported as under expressed in MCL [42] while CD1d was highly expressed [19]. It thus seems that there are some features in common between human B-CLL and MCL on the one hand and murine B-1a cells on the other.

Gene Expression of B-CLL Shows Higher Similarities to Memory B-Cells than to Naïve B-Cells or CD5+ B-Cells from Cord Blood

Approximately 60% of B-CLL has mutations in the IgVH gene and the mutational state has been found to provide new prognostic information [34, 43]. B-CLL patients without evidence of somatic mutation (>98% homology to germ line) in the highly variable regions of the immunoglobulin heavy chain genes have a more malignant course than those with mutated IgVH.

Based on recent microarray studies of gene expression, zeta associated protein-70 (ZAP-70) expression has been found to be a marker for IgVH mutational status in B-CLL [44, 45]. The tyrosine kinase ZAP-70 is part of the CD3 receptor-signaling complex in T-lymphocytes. Recent studies have shown an increased B-cell receptor signaling both in ZAP-70+, IgVH non-mutated B-CLL and also in the rare IgVH mutated cases that express ZAP-70 [46].

Among other genes that have been reported to be differentially expressed in IgVH mutated and non-mutated cases are FGFR1, gravin, dystrofin, PAK1 and BCL-7 [44] as well as PAX5 and TCF1, which are a downstream targets in the Wnt pathway [47]. PAX5 is a key regulator of B-cell differentiation (Figure 1). Thus, mice with a mutationally inactivated Pax5 gene lack B-lymphocytes [48]. In early lymphoid development, Notch and Pax5 have opposing effects, Notch expression being a signature of T-cell lineage differentiation, while Pax5 is essential for B lineage commitment and maintenance [49, 50].

It was hypothesised that B-CLL with non-mutated IgVH genes would represent malignant transformation of naïve B-cells, while Ig-gene mutated cases would stem from memory B-cells. However, results from gene expression profiling using the Affymetrix U95A chip or the Lymphochip appears not to support this hypothesis, since a large overlap in the transcriptomes between IgVH-mutated and non-mutated cases was shown [44, 51]. Importantly, both IgVH-mutated and non-mutated B-CLL have gene expression profiles that are more similar to CD27+ memory B-cells than to CD27-, naïve B-cells [51] or CD5+ B-cells isolated from cord blood [44, 51].

Gene Expression in B-Cell Compartments of Normal Lymphoid Organs Compared to MCL

Different strategies have been used to analyse gene expression in B-cell lymphomas. In most publications concerning B-CLL, tumor cells have been isolated from peripheral blood and there are only few reports of gene expression in the solid tumor counterpart, the small lymphocytic lymphoma (SLL). By the use of pure tumor cell populations subtle changes in gene expression that otherwise could have been masked by infiltrating normal cells can be detected. However, in the majority of published investigations, for other lymphoma subtypes whole tumor tissue has been used. This enables analysis also of the stromal compartment and the infiltrating normal cells. The gene expression in the tumor tissue has been compared to reactive lymphoid tissue or to purified subpopulations of B-lymphocytes. The latter approach facilitates the comparison with putative cells-of-origin. However, since the stromal compartment is not represented, important information might be lost.

An alternative approach would be to microdissect areas representing naïve, activated and memory B-cells from normal lymphoid organs in such a way that both the B-cells residing in these areas as well as the stromal compartment could be analyzed. Such an approach has recently been reported by Shen *et al.* [41]. Gene expression in different compartments of human peripheral lymphoid organs has been investigated; the germinal center, the mantle zone and the marginal zone. The mantle zone contains B-cells with non-mutated immunoglobulin genes, high expression of surface IgD and negativity for CD27. In this area generally very few T-cells reside. Most marginal zone B-cells in adults are high in expression of surface IgM, low in IgD and are CD27+ memory-like cells.

Shen et al. found that the cells in the normal marginal zone expressed specific cytokines and cytokine receptors that were not highly expressed in the other compartments [41]. There was for instance a high expression of lymphotoxin, IL-11 and receptors for IL-17 and IL-7. Furthermore, the cannabinoid receptor type 2 (CNR2) that is involved in immune regulation was highly expressed in the marginal zone but not in the other B-cell compartments investigated. In germinal centres there was a high expression of genes involved in proliferation, DNA repair and replication. Also in this compartment there was a differential expression of certain genes such as RGS13 that regulate migration and chemokine response.

While the cells in the marginal zone expressed low levels of proliferation-associated genes the mantle zone cells had highly variable expression of these genes – not as high as seen in germinal centers, but substantially higher than in marginal zones [41]. Interestingly, cyclin G2 was highly expressed in the mantle zone. This gene has been associated to suppression of proliferation [52].

How do the gene expression profiles of the microdissected normal B-cell compartments compare to what is known for mantle cell lymphoma? Mantle cell lymphoma gene expression profiling [5, 19, 53, 54] as well as other studies [55] have revealed high expression of genes involved in proliferation and cell division. In the normal mantle zone, very few proliferating cells are detected by immunohistochemical staining, while gene expression profiling has suggested that there is a slightly elevated expression of genes involved in replication and cell division when compared to the marginal zone [41]. The high expression of cyclin G2 in the mantle zone was suggested to counter-act the proliferative signals [41, 52] and it could therefore be hypothesized that dysregulated cyclin G2 could be part of the pathogenesis in mantle cell lymphoma as has been suggested for thyroid carcinoma [56]. Cyclin G2 has indeed been reported as an under-expressed gene in mantle cell lymphoma compared to sorted B-cell subpopulations [42], suggesting that dysregulation of cyclin G2 could be a new oncogenic event in MCL.

Enhanced Proliferation and Differences in Genes Involved in G-Protein Coupled Signaling in MCL

Most MCL exhibit non-mutated IgVH genes and the proportion of cases with mutated IgVH genes is lower than in B-CLL, varying between 15-30% in different recent series [57-

60]. MCL has therefore been considered to arise from naïve B-cells. In gene expression profiling of a small series of MCL a heterogenous pattern of clustering was found; with 4 out of 7 MCL clustering together with sorted populations of resting naïve or memory B-cells [42].

As in B-CLL, MCL is characterized by a biased use of IgVH genes that seems to correlate with prognosis [57-59]. There is an over-representation of the VH3-21 gene also in MCL. However, in contrast to B-CLL, in MCL these cases show no evidence of somatic mutations in the IgVH genes. Moreover, MCL cases using VH3-21 are associated with a better prognosis. As of today there is limited information on differences in gene expression associated with mutational status of the Ig receptor in MCL. However, in a recent report a few genes such as EGR2, SMARCA5 and replication protein A1 were found to be differently expressed in mutated cases [61].

In MCL there is generally higher expression of genes involved in proliferation compared to B-CLL/SLL [5, 19, 51, 55]. In addition to cyclin D1, some mantle cell lymphomas express high levels of genes such as topoisomerase IIA, CDK4, PCNA, BUB1. These genes are highly expressed in germinal centres in normal lymph nodes, but in the context of MCL they characterize the highly proliferative tumors with adverse prognosis [5, 62]. We have shown that many of these highly proliferative MCL cases express a more stable, truncated cyclin D1 transcript, lacking the 3'UTR region of the mRNA [5, 19] [62]. Further studies of this MCL variant has revealed that the short, stable cyclin D1 mRNA isoform is associated with 3.4 times higher levels of cyclin D1 mRNA coding region and higher cyclin D1 protein levels [62]. These MCL tumors have high expression of a number of cell cycle regulating genes such as cyclin A1 and cyclin B1 leading to further disturbances of cell cycle regulation, increased cell proliferation and a more aggressive disease.

We have recently shown that genes encoding proteins involved in G-protein mediated signaling are differentially regulated in MCL compared to reactive lymphoid tissue [19]. Regulator of G protein signaling 13 (RGS13) and other RGS proteins have an important role in regulating chemokine-induced cell migration and germinal center reactions [63-65]. Especially RGS13 is localized to germinal centers [63] and was absent in mantle cell lymphoma while highly expressed in reactive tonsils and lymph nodes [19]. By gene expression profiling we and others found that the G-protein coupled receptor, cannabinoid receptor type I (CB1), which is highly expressed in brain and is involved in regulation of food-intake, is significantly expressed in MCL [19, 42]. Recent data suggest that the CB1 gene also is expressed in B-CLL with 11q22-q23 deletions [27]. Also the cannabinoid receptor 2 is highly expressed in MCL [19]. In normal lymphoid tissue this gene seems to be preferentially expressed in the mantle/marginal zones [41, 66, 67] and has immunoregulatory functions. We have investigated the functional importance of high cannabinoid receptor expression in MCL and found that targeting of cannabinoid receptors impair proliferation and induce cell death [68] in MCL. The cannabinoid system could thus be considered as a new therapeutic target, originally defined by array screening of the mRNA pool.

Marginal Zone Lymphoma is a Heterogenous Tumor that Can Be Distinguished from other Small Cell Lymphomas by Gene Expression Profiling

The lymphoma entity, marginal zone lymphoma (MZL), seem to contain several different subtypes with different genomic alterations, some of which are preferentially seen in MZL located at distinct anatomical sites [69-74]. While most MZL exhibit somatic mutations of IgVH genes, non-mutated cases are associated with a worse prognosis [75].

The MZL entity displays a highly variable immunohistochemical profile [76] that in some instances overlaps with other lymphoma entities. It is therefore very interesting that by gene expression profiling a small set of genes could be used to differentiate MZL from CLL or MCL in a study containing 37 marginal zone lymphomas [77]. In addition to a large MZL cluster corresponding to genes characteristic for "spleen" there was another MZL cluster specifically associated with the marginal zone signature [77] irrespectively of which organ was affected by the lymphoma. This cluster contained genes involved in intracellular signaling or transcription such as AKT1 and the transcription factor TPCP2. Also genes for many S100 proteins were highly expressed.

Apart from this study by Thieblemont *et al.* [77] not much is known concerning global gene expression profiles in MZL. In this study, as few as 44 genes grouped in 4 clusters could differentiate between SLL, MCL and MZL with a high degree of certainty. Although these findings have to be confirmed in other studies, the results are very promising indeed and clearly indicates that microarray analysis might have an important role not only in research but also in diagnostic practice.

Primary Immunodeficiency Disorders

B-lymphocytes at different developmental stages express distinct sets of important signaling components. Diseases associated with constitutively active, non-functional states of these genes highlight their importance. This is true not only for lymphomas but also for PIDs. In the decade that has passed since the molecular basis of XLA was identified, mutated genes in more than 100 PIDs have been described in the literature.

The PIDs are characterized by defects of specific functions of the immune system, most often manifested by increased susceptibility to infections. The disorders, although rare, provide a wide spectrum of defects, including antibody deficiencies, cellular immune deficiencies, combined immune deficiencies, phagocytic defects and complement defects [78-83], where antibody deficiencies represent the most common types of PIDs in humans [84].

These disorders involve components of both the adaptive and innate immune system, including T- and B-lymphocytes, NK cells, phagocytic cells and various complement proteins. In general, the T-cell deficiencies are clinically more severe than those associated with antibody deficiency disorders. In the absence of treatment, those affected by defective cellular immunity, rarely survive beyond infancy or childhood. As with the T-cell immune deficiencies, the spectrum of antibody deficiency disorders is broad, ranging from the most severe type, with next to absent B-cells and immunoglobulins, to patients having selective Ig isotype defects, to those with almost normal immunoglobulins in the serum.

In B-cell deficiencies, the symptoms often do not appear until 9-12 months of age, since the patient has acquired protective levels of Ig from the mother. In addition to a high susceptibility to infections certain antibody deficiency disorders are associated with other features in their clinical presentation, e.g. autoimmunity and malignancies. Many of the genes involved in antibody deficiencies have now been identified and this has led to a better understanding and elucidation of the B-lymphocyte development and the differentiation pathways. However, it is important to consider the various stages of B-lymphocyte differentiation and development in order to understand the abnormal mechanisms underlying the different antibody immunodeficiencies. PIDs affect several subsets of cells. In Figure 1 the effects on B-cell development and differentiation is shown. This figure also summarizes the developmental stages where different immunodeficiencies and lymphomas are suggested to occur.

XLA (X-Linked Agammaglobulinemia)

X-linked agammaglobulinemia (XLA) was reported in 1952 by Dr. Ogden C. Bruton [80]. In 1993, two independent laboratories cloned the gene responsible for XLA [85, 86]. The gene was named Bruton tyrosine kinase (*BTK*) after Dr. Bruton's discovery of the disease. XLA patients present a vast number of distinct mutations in the *BTK* gene [78, 87, 88] all of which affects their B-cell development and differentiation [89, 90]. For several years the existence of a genotype-phenotype correlation remained elusive. However, it was recently demonstrated that there in fact exists such a correlation, since missense mutations are highly restricted in XLA with only certain residues being sensitive [91]. This phenomenon is particularly evident for CpG sites, which are frequently mutated and where certain codons cause XLA upon non-sense mutations, while amino acid substitutions seem to be tolerated. It is likely that certain forms of splice site mutations result in a milder form of the disease. However, such patients have not been described to date.

Approximately 1 in every 380,000 males is affected with XLA [92]. The onset of infections occurs during the later part of the first year of life, when maternal antibodies are reduced below protective levels [83, 93]. There is an almost total absence, or a marked deficiency, of serum Igs in patients affected with XLA. The percentage of circulating B-cells is reduced or extremely low. However, pro-B-cells numbers in the bone marrow are normal or even increased in number [89, 90]. The defect is B-cell intrinsic not affecting T-lymphocytes and other lymphoid subpopulations, which appear in normal numbers. Lymphoid tissues show absence of plasma cells, lymphoid follicles and germinal centers. The treatment for XLA remains the same as the one Dr. Bruton himself introduced in the 1950s, a combination of gammaglobulin substitution and antibiotics and most patients have few severe infections with this regular treatment.

SIgAD (Selective IgA Deficiency)

SIgAD is the most common form of PID in the Western world and affects approximately 1 in every 600 individuals [84]. Patients affected with SIgAD have markedly reduced or absent serum IgA levels but normal IgM and IgG concentrations. Many individuals with SIgAD are clinically asymptomatic but those patients with symptoms display respiratory tract infections and involvement of the gastrointestinal tract with giardiasis and nodular lymphoid hyperplasia [94]. The genes encoding the α heavy chain genes are intact [95], whereas abnormalities lies in isotype switching have been found [96]. A number of abnormalities in the cytokine network have been observed in SIgAD but no mutation in cytokine genes have been found to date [97]. For this reason a relevant animal model has not been created for SIgAD and therefore primary peripheral blood B-cells or Epstein-Barr Virus (EBV) transformed B-cells from patients are commonly used as the research tool in molecular biology.

CVID (Common Variable Immunodeficiency)

Common variable immunodeficiency (CVID), also called acquired hypogammaglobulinemia, adult-onset hypogammaglobulinemia or dysgammaglobulinemia, is a heterogenous group of disorders involving both B- and T-cell immune function. Patients present with recurrent bacterial infections, decreased serum Ig levels and deficient antibody production. The wide spectrum of CVID phenotypes is also reflected in the age of onset of the disease, which ranges from early childhood to young adults with an average age of onset of around 25 years. Variability is also found in the degree and type of hypogammaglobulinemia and the patients may also suffer from chronic inflammatory and autoimmune conditions [98]. Several mechanisms have been proposed to explain the immune abnormalities in patients with CVID such as B-cell defects, excessive T-suppressor cell activity [99], deficient T-cell helper function [100], cytokine deficiency [101, 102] and suboptimal T- and B-cell interactions through the CD40 ligand [103]. The clinical variability of CVID is likely to be a reflection of a complex multigene involvement of the disease. The concept that more than one gene is responsible for the disease is supported by the involvement of one or more genes within the MHC region on chromosome 6 in the pathogenesis for both SIgAD and CVID [104-106]. Recently, mutations in the ICOS as well as the TACI genes were found to cause a rare form of CVID [107, 108].

ICF (Immunodeficiency, Centromeric Region Instability and Facial Anomalies)

ICF is an autosomal recessive disease caused by mutations in the DNA methyltransferase 3B gene (*DNMT3B*). The disease is a rare chromosome instability syndrome and is the only disease shown to be associated with mutations in a DNA methyltransferase gene [109]. In this syndrome the immunodeficiency is characterized by agammaglobulinemia in the presence of B-cells. DNA rearrangements are targeted to the centromere-adjacent heterochromatic region of chromosomes 1, 16, and sometimes 9 in mitogen-stimulated lymphocytes. Immune-related problems in ICF patients are generally, but not always, severe. Many patients exhibit low *in vitro* stimulation indices for B- or T-cells upon exposure to mitogens or antigens. ICF is usually diagnosed in the early years [110, 111], and these patients often die in infancy or childhood [110, 112-114]. The cause of death in these patients is usually infection.

Gene Expression Profiling in PIDs

With regards to immunodeficiency diseases, analysis with the microarray technology has been conducted in two ways. First, it has been used to study gene expression in patients with known defects in order to better understand the basic biology and pathophysiology of the disease [7, 8, 115]. Second, microarray analysis can be used as an exploratory tool to study differential gene expression in immunodeficiency diseases of unknown cause, such as CVID or SIgAD [9].

Depending on the type of immunodeficiency disease, patients may have low numbers of circulating lymphocytes. An example is XLA, where the patients essentially lack antibody producing B-lymphocytes. This makes it hard, next to impossible, to do research on primary B-cells from XLA patients. A way to circumvent this is to immortalize B-cells from these individuals by EBV transformation. In CVID and SIgAD, where circulating B-cells usually are found, peripheral blood mononuclear cells have been used for gene expression profiling.

Both cDNA spotted filter arrays as well as Affymetrix™ high-density oligonucleotide arrays have been used in order to study the influence of Btk in cellular signaling [7]. EBV transformed B-cells from healthy individuals were used in comparison to XLA patients. The profile revealed a number of genes differentially expressed between healthy controls and patients. These genes included those expressed in haematopoietic cells including genes from groups corresponding to receptors, kinases, transcription factors and cofactors. Several interesting genes such as *FYN*, *HCK* and *CYP1B1* were more highly expressed in the XLA patient compared to the normal control. One EST and eight complementary DNA clones with unknown function were down-regulated in the XLA patient cell line. In addition to these results, the B-cell specific OCT binding factor-1 was found to be up-regulated in the cell lines from the XLA patient in comparison to the healthy control. The study clearly demonstrated that the influence of Btk mutations can be detected by the use of microarrays, and that this type of analysis can infer potential roles for genes with both unknown and known function. Furthermore, primary splenic B-cells from *Xid* and Btk KO mice were analyzed and compared to normal splenic B-cells from healthy mice of the same genetic background [2]. The authors found approximately 40 genes differing between the Btk defective B-cells compared to healthy controls. Amongst these 40 genes, microsomal epoxide hydrolase (mEH), CD9 and ionized calcium-binding adaptor molecule 1 (Iba1) were found to be affected by the Btk defect.

Other immunodeficiency disorders that have been studied using gene expression profiling are the related immunodeficiency diseases, CVID and SIgAD, as well as the rare ICF syndrome. [109, 116, 117]. In a recent study Qin *et. al.* show [9] that the gene expression pattern from both the CVID and the SIgAD patients gave similar results with some skew-ness towards the Th1 pathway. The results found in both patient groups were in agreement with the observed clinical phenotype. Gene expression microarrays have also been used to analyse EBV transformed B-cell lines from ICF patients [8]. The aim was to identify dysregulation of gene expression that might explain the symptoms of ICF. Genes, such as *RGS1*, *GU-CY1B3* and *F13A1*, which are known to play an important role in the immune function, were identified. The result also suggests that mutations in the gene *DNMT3B* cause lymphogenesis-associated gene dysregulation by indirect effects on gene expression that interfere with normal lymphocyte signaling, maturation and migration.

Conclusion

During development cells differentiate from precursors, starting with the fertilized egg. Numerous markers, such as cell surface proteins, characterize different developmental stages. These processes have best been studied in model organisms with *C. elegans* being the prototype for multicellular species. The adult worm consists of 959 somatic cells compared to approximately 10^{14} in humans. It is considered that *C. elegans* is formed by 10 to 12 different cell types and the corresponding number in humans has been estimated to roughly 250. However, a key issue is what distinguishes one cell type from another.

Analysis of gene transcription of B cell lymphomas and PID has widened our understanding of normal B cell differentiation in addition to identifying new disease entities. Array based expression analysis will certainly continue to provide crucial information in both clinical and experimental medicine. Identifying specific transcriptional profiles in malignant lymphomas can also be used to better define from which normal B-cell subset the malignant cells have originated and thus provide new insights in the pathogenesis of the disease. Using transcriptome and proteome analysis in combination with more assays of gene function will allow more precise definition of developmental stages with important biological characteristics.

Acknowledgements

This work was supported by the Swedish Science Council, the Wallenberg foundation, the Swedish Cancer Foundation, the Magnus Bergvalls foundation, European union grant EURO-POLICY-PID and the Cancer Society of Stockholm.

REFERENCES

[1] Consortium, I.H.G.S., *Finishing the euchromatic sequence of the human genome.* Nature, 2004. 431(7011): p. 931-45.

[2] Lindvall, J.M., et al., *Gene expression profile of B cells from Xid mice and Btk knockout mice.* Eur J Immunol, 2004. 34(7): p. 1981-91.

[3] Oltz, E.M., et al., *A novel regulatory myosin light chain gene distinguishes pre-B cell subsets and is IL-7 inducible.* Embo J, 1992. 11(7): p. 2759-67.

[4] Alizadeh, A.A., et al., *Distinct types of diffuse large B-cell lymphoma identified by gene expression profiling.* Nature, 2000. 403(6769): p. 503-11.

[5] Rosenwald, A., et al., *The proliferation gene expression signature is a quantitative integrator of oncogenic events that predicts survival in mantle cell lymphoma.* Cancer Cell, 2003. 3(2): p. 185-97.

[6] Fruman, D.A., et al., *Phosphoinositide 3-kinase and Bruton's tyrosine kinase regulate overlapping sets of genes in B lymphocytes.* Proc Natl Acad Sci U S A, 2002. 99(1): p. 359-64.

[7] Islam, T.C., et al., *Expression profiling in transformed human B cells: influence of Btk mutations and comparison to B cell lymphomas using filter and oligonucleotide arrays.* Eur J Immunol, 2002. 32(4): p. 982-93.

[8] Ehrlich, M., et al., *DNA methyltransferase 3B mutations linked to the ICF syndrome cause dysregulation of lymphogenesis genes.* Hum Mol Genet, 2001. 10(25): p. 2917-31.

[9] Qin, H., et al., *Tracking gene expression in primary immunodeficiencies.* Curr Opin Allergy Clin Immunol, 2003. 3(6): p. 437-42.

[10] Dadi, H.K., A.J. Simon, and C.M. Roifman, *Effect of CD3delta deficiency on maturation of alpha/beta and gamma/delta T-cell lineages in severe combined immunodeficiency.* N Engl J Med, 2003. 349(19): p. 1821-8.

[11] Reth, M. and J. Wienands, *Initiation and processing of signals from the B cell antigen receptor.* Annu Rev Immunol, 1997. 15: p. 453-79.

[12] Chan, F.K., et al., *A domain in TNF receptors that mediates ligand-independent receptor assembly and signaling.* Science, 2000. 288(5475): p. 2351-4.

[13] Livnah, O., et al., *Crystallographic evidence for preformed dimers of erythropoietin receptor before ligand activation.* Science, 1999. 283(5404): p. 987-90.

[14] Schamel, W.W. and M. Reth, *Monomeric and oligomeric complexes of the B cell antigen receptor.* Immunity, 2000. 13(1): p. 5-14.

[15] Gold, M.R., *To make antibodies or not: signaling by the B-cell antigen receptor.* Trends Pharmacol Sci, 2002. 23(7): p. 316-24.

[16] Dal Porto, J.M., et al., *B cell antigen receptor signaling 101.* Mol Immunol, 2004. 41(6-7): p. 599-613.

[17] Clarke, P.A., R.T. Poele, and P. Workman, *Gene expression microarray technologies in the development of new therapeutic agents.* Eur J Cancer, 2004. 40(17): p. 2560-91.

[18] Villeneuve, D.J. and A.M. Parissenti, *The use of DNA microarrays to investigate the pharmacogenomics of drug response in living systems.* Curr Top Med Chem, 2004. 4(13): p. 1329-45.

[19] Islam, T.C., et al., *High level of cannabinoid receptor 1, absence of regulator of G protein signalling 13 and differential expression of Cyclin D1 in mantle cell lymphoma.* Leukemia, 2003. 17(9): p. 1880-90.

[20] William, J., et al., *Evolution of autoantibody responses via somatic hypermutation outside of germinal centers.* Science, 2002. 297(5589): p. 2066-70.

[21] Weller, S., et al., *CD40-CD40L independent Ig gene hypermutation suggests a second B cell diversification pathway in humans.* Proc Natl Acad Sci U S A, 2001. 98(3): p. 1166-70.

[22] Welzel, N., et al., *Templated nucleotide addition and immunoglobulin JH-gene utilization in t(11;14) junctions: implications for the mechanism of translocation and the origin of mantle cell lymphoma.* Cancer Res, 2001. 61(4): p. 1629-36.

[23] Stamatopoulos, K., et al., *Molecular analysis of bcl-1/IgH junctional sequences in mantle cell lymphoma: potential mechanism of the t(11;14) chromosomal translocation.* Br J Haematol, 1999. 105(1): p. 190-7.

[24] Stilgenbauer, S., et al., *Genetics of chronic lymphocytic leukemia: genomic aberrations and V(H) gene mutation status in pathogenesis and clinical course.* Leukemia, 2002. 16(6): p. 993-1007.

[25] Fegan, C., et al., *Karyotypic evolution in CLL: identification of a new sub-group of patients with deletions of 11q and advanced or progressive disease.* Leukemia, 1995. 9(12): p. 2003-8.

[26] Juliusson, G., et al., *Prognostic information from cytogenetic analysis in chronic B-lymphocytic leukemia and leukemic immunocytoma.* Blood, 1985. 65(1): p. 134-41.

[27] Haslinger, C., et al., *Microarray gene expression profiling of B-cell chronic lymphocytic leukemia subgroups defined by genomic aberrations and VH mutation status.* J Clin Oncol, 2004. 22(19): p. 3937-49.

[28] Berland, R. and H.H. Wortis, *Origins and functions of B-1 cells with notes on the role of CD5.* Annu Rev Immunol, 2002. 20: p. 253-300.

[29] Martin, F. and J.F. Kearney, *B1 cells: similarities and differences with other B cell subsets.* Curr Opin Immunol, 2001. 13(2): p. 195-201.

[30] Hardy, R.R., C.J. Wei, and K. Hayakawa, *Selection during development of VH11+ B cells: a model for natural autoantibody-producing CD5+ B cells.* Immunol Rev, 2004. 197: p. 60-74.

[31] Hayakawa, K., et al., *Positive selection of anti-thy-1 autoreactive B-1 cells and natural serum autoantibody production independent from bone marrow B cell development.* J Exp Med, 2003. 197(1): p. 87-99.

[32] Kipps, T.J., et al., *Developmentally restricted immunoglobulin heavy chain variable region gene expressed at high frequency in chronic lymphocytic leukemia.* Proc Natl Acad Sci U S A, 1989. 86(15): p. 5913-7.

[33] Fais, F., et al., *Chronic lymphocytic leukemia B cells express restricted sets of mutated and unmutated antigen receptors.* J Clin Invest, 1998. 102(8): p. 1515-25.

[34] Hamblin, T.J., et al., *Unmutated Ig V(H) genes are associated with a more aggressive form of chronic lymphocytic leukemia.* Blood, 1999. 94(6): p. 1848-54.

[35] Tobin, G., et al., *Somatically mutated Ig V(H)3-21 genes characterize a new subset of chronic lymphocytic leukemia.* Blood, 2002. 99(6): p. 2262-4.

[36] Tobin, G., et al., *Subsets with restricted immunoglobulin gene rearrangement features indicate a role for antigen selection in the development of chronic lymphocytic leukemia.* Blood, 2004.

[37] Ghia, P., et al., *Geographical patterns and pathogenetic implications of IGHV gene usage in chronic lymphocytic leukemia: the lesson of the IGHV3-21 gene.* Blood, 2004.

[38] Messmer, B.T., et al., *Multiple distinct sets of stereotyped antigen receptors indicate a role for antigen in promoting chronic lymphocytic leukemia.* J Exp Med, 2004. 200(4): p. 519-25.

[39] Weller, S., et al., *Human blood IgM "memory" B cells are circulating splenic marginal zone B cells harboring a pre-diversified immunoglobulin repertoire.* Blood, 2004.

[40] Kretschmer, K., et al., *Antibody repertoire and gene expression profile: implications for different developmental and functional traits of splenic and peritoneal B-1 lymphocytes.* J Immunol, 2003. 171(3): p. 1192-201.

[41] Shen, Y., et al., *Distinct gene expression profiles in different B-cell compartments in human peripheral lymphoid organs.* BMC Immunol, 2004. 5(1): p. 20.

[42] Ek, S., et al., *Mantle cell lymphomas express a distinct genetic signature affecting lymphocyte trafficking and growth regulation as compared with subpopulations of normal human B cells.* Cancer Res, 2002. 62(15): p. 4398-405.

[43] Damle, R.N., et al., *Ig V gene mutation status and CD38 expression as novel prognostic indicators in chronic lymphocytic leukemia.* Blood, 1999. 94(6): p. 1840-7.

[44] Rosenwald, A., et al., *Relation of gene expression phenotype to immunoglobulin mutation genotype in B cell chronic lymphocytic leukemia.* J Exp Med, 2001. 194(11): p. 1639-47.

[45] Wiestner, A., et al., *ZAP-70 expression identifies a chronic lymphocytic leukemia subtype with unmutated immunoglobulin genes, inferior clinical outcome, and distinct gene expression profile.* Blood, 2003. 101(12): p. 4944-51.

[46] Chen, L., et al., *Expression of ZAP-70 is associated with increased B-cell receptor signaling in chronic lymphocytic leukemia.* Blood, 2002. 100(13): p. 4609-14.

[47] Ferrer, A., et al., *Different gene expression in immunoglobulin-mutated and immunoglobulin-unmutated forms of chronic lymphocytic leukemia.* Cancer Genet Cytogenet, 2004. 153(1): p. 69-72.

[48] Busslinger, M., *Transcriptional control of early B cell development.* Annu Rev Immunol, 2004. 22: p. 55-79.

[49] Hoflinger, S., et al., *Analysis of Notch1 function by in vitro T cell differentiation of Pax5 mutant lymphoid progenitors.* J Immunol, 2004. 173(6): p. 3935-44.

[50] Tagoh, H., et al., *Epigenetic silencing of the c-fms locus during B-lymphopoiesis occurs in discrete steps and is reversible.* Embo J, 2004. 23(21): p. 4275-85.

[51] Klein, U., et al., *Gene Expression Profiling of B Cell Chronic Lymphocytic Leukemia Reveals a Homogeneous Phenotype Related to Memory B Cells.* J. Exp. Med., 2001. 194(11): p. 1625-1638.

[52] Horne, M.C., et al., *Cyclin G2 is up-regulated during growth inhibition and B cell antigen receptor-mediated cell cycle arrest.* J Biol Chem, 1997. 272(19): p. 12650-61.

[53] Ek, S., et al., *Increased expression of Ki-67 in mantle cell lymphoma is associated with de-regulation of several cell cycle regulatory components, as identified by global gene expression analysis.* Haematologica, 2004. 89(6): p. 686-95.

[54] Schmechel, S.C., et al., *Identification of genes whose expression patterns differ in benign lymphoid tissue and follicular, mantle cell, and small lymphocytic lymphoma.* Leukemia, 2004. 18(4): p. 841-55.

[55] Korz, C., et al., *Evidence for distinct pathomechanisms in B-cell chronic lymphocytic leukemia and mantle cell lymphoma by quantitative expression analysis of cell cycle and apoptosis-associated genes.* Blood, 2002. 99(12): p. 4554-61.

[56] Ito, Y., et al., *Decreased expression of cyclin G2 is significantly linked to the malignant transformation of papillary carcinoma of the thyroid.* Anticancer Res, 2003. 23(3B): p. 2335-8.

[57] Walsh, S.H., et al., *Mutated VH genes and preferential VH3-21 use define new subsets of mantle cell lymphoma.* Blood, 2003. 101(10): p. 4047-54.

[58] Kienle, D., et al., *VH mutation status and VDJ rearrangement structure in mantle cell lymphoma: correlation with genomic aberrations, clinical characteristics, and outcome.* Blood, 2003. 102(8): p. 3003-9.

[59] Camacho, F.I., et al., *Molecular heterogeneity in MCL defined by the use of specific VH genes and the frequency of somatic mutations.* Blood, 2003. 101(10): p. 4042-6.

[60] Orchard, J., et al., *A subset of t(11;14) lymphoma with mantle cell features displays mutated IgVH genes and includes patients with good prognosis, nonnodal disease.* Blood, 2003. 101(12): p. 4975-81.

[61] Martinez, N., et al., *The molecular signature of mantle cell lymphoma reveals multiple signals favoring cell survival.* Cancer Res, 2003. 63(23): p. 8226-32.

[62] Sander, B., et al., *Mantle cell lymphomas with low levels of cyclin D1 long mRNA transcripts are highly proliferative and can be discriminated by elevated cyclin A2 and cyclin B1.* Int. J. Cancer, 2005. 117(3): p. 418-30.

[63] Shi, G.X., et al., *RGS13 regulates germinal center B lymphocytes responsiveness to CXC chemokine ligand (CXCL)12 and CXCL13.* J Immunol, 2002. 169(5): p. 2507-15.

[64] Shi, G.X., et al., *Toll-like receptor signaling alters the expression of regulator of G protein signaling proteins in dendritic cells: implications for G protein-coupled receptor signaling.* J Immunol, 2004. 172(9): p. 5175-84.

[65] Moratz, C., K. Harrison, and J.H. Kehrl, *Regulation of Chemokine-Induced Lymphocyte Migration by RGS Proteins.* Methods Enzymol, 2004. 389: p. 15-32.

[66] Rayman, N., et al., *Distinct expression profiles of the peripheral cannabinoid receptor in lymphoid tissues depending on receptor activation status.* J Immunol, 2004. 172(4): p. 2111-7.

[67] Galiegue, S., et al., *Expression of central and peripheral cannabinoid receptors in human immune tissues and leukocyte subpopulations.* Eur J Biochem, 1995. 232(1): p. 54-61.

[68] Flygare J, G.K., Kimby E, Christensson B, Sander B., *Cannabinoid receptor ligands mediate growth inhibition and cell death in mantle cell lymphoma.* FEBS Lett, 2005. 579(30): p. 6885-9.

[69] Sanchez-Izquierdo, D., et al., *MALT1 is deregulated by both chromosomal translocation and amplification in B-cell non-Hodgkin lymphoma.* Blood, 2003. 101(11): p. 4539-46.

[70] Remstein, E.D., et al., *Primary pulmonary MALT lymphomas show frequent and heterogeneous cytogenetic abnormalities, including aneuploidy and translocations involving API2 and MALT1 and IGH and MALT1.* Leukemia, 2004. 18(1): p. 156-60.

[71] Streubel, B., et al., *T(14;18)(q32;q21) involving IGH and MALT1 is a frequent chromosomal aberration in MALT lymphoma.* Blood, 2003. 101(6): p. 2335-9.

[72] Morgan, J.A., et al., *Breakpoints of the t(11;18)(q21;q21) in mucosa-associated lymphoid tissue (MALT) lymphoma lie within or near the previously undescribed gene MALT1 in chromosome 18.* Cancer Res, 1999. 59(24): p. 6205-13.

[73] Dierlamm, J., et al., *The apoptosis inhibitor gene API2 and a novel 18q gene, MLT, are recurrently rearranged in the t(11;18)(q21;q21)p6ssociated with mucosa-associated lymphoid tissue lymphomas.* Blood, 1999. 93(11): p. 3601-9.

[74] Akagi, T., et al., *A novel gene, MALT1 at 18q21, is involved in t(11;18) (q21;q21) found in low-grade B-cell lymphoma of mucosa-associated lymphoid tissue.* Oncogene, 1999. 18(42): p. 5785-94.

[75] Algara, P., et al., *Analysis of the IgV(H) somatic mutations in splenic marginal zone lymphoma defines a group of unmutated cases with frequent 7q deletion and adverse clinical course.* Blood, 2002. 99(4): p. 1299-304.

[76] Camacho, F.I., et al., *Nodal marginal zone lymphoma: a heterogeneous tumor: a comprehensive analysis of a series of 27 cases.* Am J Surg Pathol, 2003. 27(6): p. 762-71.

[77] Thieblemont, C., et al., *Small lymphocytic lymphoma, marginal zone B-cell lymphoma, and mantle cell lymphoma exhibit distinct gene-expression profiles allowing molecular diagnosis.* Blood, 2004. 103(7): p. 2727-37.

[78] Vihinen, M., et al., *Mutations of the human BTK gene coding for bruton tyrosine kinase in X-linked agammaglobulinemia.* Hum Mutat, 1999. 13(4): p. 280-5.

[79] Bonilla, F.A. and R.S. Geha, *12. Primary immunodeficiency diseases.* J Allergy Clin Immunol, 2003. 111(2 Suppl): p. S571-81.

[80] Bruton, O.C., *Agammaglobulinemia.* Pediatrics, 1952. 9: p. 722-27.

[81] Buckley, R.H., *Primary cellular immunodeficiencies.* J Allergy Clin Immunol, 2002. 109(5): p. 747-57.

[82] Chinen, J. and W.T. Shearer, *Basic and clinical immunology.* J Allergy Clin Immunol, 2003. 111(3 Suppl): p. S813-8.

[83] Ochs, H.D. and C.I. Smith, *X-linked agammaglobulinemia. A clinical and molecular analysis.* Medicine (Baltimore), 1996. 75(6): p. 287-99.

[84] Ochs, H.D., C.I.E. Smith, and J.M. Puck, *Primary Immunodeficiency Diseases: A Molecular and Genetic Approach.* 1998: Oxford University Press.

[85] Vetrie, D., *Isolation of the defective gene in X linked agammaglobulinaemia.* J Med Genet, 1993. 30(6): p. 452-3.

[86] Tsukada, S., et al., *Deficient expression of a B cell cytoplasmic tyrosine kinase in human X-linked agammaglobulinemia.* Cell, 1993. 72(2): p. 279-90.

[87] Vihinen, M., et al., *BTKbase: a database of XLA-causing mutations. International Study Group.* Immunol Today, 1995. 16(10): p. 460-5.
[88] Vihinen, M., et al., *BTKbase, mutation database for X-linked agammaglobulinemia (XLA).* Nucleic Acids Res, 1997. 25(1): p. 166-71.
[89] Campana, D., et al., *Phenotypic features and proliferative activity of B cell progenitors in X-linked agammaglobulinemia.* J Immunol, 1990. 145(6): p. 1675-80.
[90] Noordzij, J.G., et al., *Composition of precursor B-cell compartment in bone marrow from patients with X-linked agammaglobulinemia compared with healthy children.* Pediatr Res, 2002. 51(2): p. 159-68.
[91] Lindvall, J.M., et al., *Bruton's tyrosine kinase: cell biology, sequence conservation, mutation spectrum, siRNA modifications, and expression profiling.* Immunol Rev, 2005. 203: p. 200-15.
[92] Winkelstein, J.A., et al., *X-linked agammaglobulinemia: report on a United States registry of 201 patients.* Medicine (Baltimore), 2006. 85(4): p. 193-202.
[93] Sideras, P. and C.I. Smith, *Molecular and cellular aspects of X-linked agammaglobulinemia.* Adv Immunol, 1995. 59: p. 135-223.
[94] Gryboski, J.D., et al., *Selective immunoglobulin A deficiency and intestinal nodular lymphoid hyperplasia: correction of diarrhea with antibiotics and plasma.* Pediatrics, 1968. 42(5): p. 833-7.
[95] Hammarstrom, L., G.G. de Lange, and C.I. Smith, *IgA2 allotypes determined by restriction fragment length polymorphism in IgA deficiency. Re-expression of the silent A2m(2) allotype in the children of IgA-deficient patients.* J Immunogenet, 1987. 14(4-5): p. 197-201.
[96] Islam, K.B., et al., *Molecular analysis of IgA deficiency. Evidence for impaired switching to IgA.* J Immunol, 1994. 152(3): p. 1442-52.
[97] Briere, F., et al., *Interleukin 10 induces B lymphocytes from IgA-deficient patients to secrete IgA.* J Clin Invest, 1994. 94(1): p. 97-104.
[98] Cunningham-Rundles, C. and C. Bodian, *Common variable immunodeficiency: clinical and immunological features of 248 patients.* Clin Immunol, 1999. 92(1): p. 34-48.
[99] Waldmann, T.A., et al., *Role of suppressor T cells in pathogenesis of common variable hypogammaglobulinaemia.* Lancet, 1974. 2(7881): p. 609-13.
[100] Reinherz, E.L., et al., *Immunodeficiency associated with loss of T4+ inducer T-cell function.* N Engl J Med, 1981. 304(14): p. 811-6.
[101] Eisenstein, E.M., K. Chua, and W. Strober, *B cell differentiation defects in common variable immunodeficiency are ameliorated after stimulation with anti-CD40 antibody and IL-10.* J Immunol, 1994. 152(12): p. 5957-68.
[102] Sneller, M.C. and W. Strober, *Abnormalities of lymphokine gene expression in patients with common variable immunodeficiency.* J Immunol, 1990. 144(10): p. 3762-9.
[103] Farrington, M., et al., *CD40 ligand expression is defective in a subset of patients with common variable immunodeficiency.* Proc Natl Acad Sci U S A, 1994. 91(3): p. 1099-103.
[104] Ashman, R.F., et al., *Genetic and immunologic analysis of a family containing five patients with common-variable immune deficiency or selective IgA deficiency.* J Clin Immunol, 1992. 12(6): p. 406-14.

[105] Olerup, O., C.I. Smith, and L. Hammarstrom, *Different amino acids at position 57 of the HLA-DQ beta chain associated with susceptibility and resistance to IgA deficiency.* Nature, 1990. 347(6290): p. 289-90.

[106] Olerup, O., et al., *Shared HLA class II-associated genetic susceptibility and resistance, related to the HLA-DQB1 gene, in IgA deficiency and common variable immunodeficiency.* Proc Natl Acad Sci U S A, 1992. 89(22): p. 10653-7.

[107] Grimbacher, B., et al., *Homozygous loss of ICOS is associated with adult-onset common variable immunodeficiency.* Nat Immunol, 2003. 4(3): p. 261-8.

[108] Salzer, U., et al., *Mutations in TNFRSF13B encoding TACI are associated with common variable immunodeficiency in humans.* Nat Genet, 2005. 37(8): p. 820-8.

[109] Xu, G.L., et al., *Chromosome instability and immunodeficiency syndrome caused by mutations in a DNA methyltransferase gene.* Nature, 1999. 402(6758): p. 187-91.

[110] De Ravel, T.J., et al., *The ICF syndrome: new case and update.* Genet Couns, 2001. 12(4): p. 379-85.

[111] Smeets, D.F., et al., *ICF syndrome: a new case and review of the literature.* Hum Genet, 1994. 94(3): p. 240-6.

[112] Sawyer, J.R., et al., *Chromosome instability in ICF syndrome: formation of micronuclei from multibranched chromosomes 1 demonstrated by fluorescence in situ hybridization.* Am J Med Genet, 1995. 56(2): p. 203-9.

[113] Fasth, A., et al., *Fragility of the centromeric region of chromosome 1 associated with combined immunodeficiency in siblings. A recessively inherited entity?* Acta Paediatr Scand, 1990. 79(6-7): p. 605-12.

[114] Brown, D.C., et al., *ICF syndrome (immunodeficiency, centromeric instability and facial anomalies): investigation of heterochromatin abnormalities and review of clinical outcome.* Hum Genet, 1995. 96(4): p. 411-6.

[115] Feske, S., et al., *Gene regulation mediated by calcium signals in T lymphocytes.* Nat Immunol, 2001. 2(4): p. 316-24.

[116] Cunningham-Rundles, C., et al., *Lymphomas of mucosal-associated lymphoid tissue in common variable immunodeficiency.* Am J Hematol, 2002. 69(3): p. 171-8.

[117] Hammarstrom, L., I. Vorechovsky, and D. Webster, *Selective IgA deficiency (SIgAD) and common variable immunodeficiency (CVID).* Clin Exp Immunol, 2000. 120(2): p. 225-31.

In: New Messenger RNA Research Communications
Editor: Lee B. Kwang, pp. 25-42
ISBN 978-1-60021-488-2

Chapter II

Messenger "RNA" as an Intercellular Signaling Molecule

Chunlin Shi, Xiong-Yan Chen and Jae-Yean Kim*
Division of Applied Life Science (BK21 program), EBRC-PMBBRC
Gyeongsang National University, Jinju 660-701, Korea

ABSTRACT

Intercellular communication is a critical biological process for the orchestration of position-dependent cell differentiation, organ development and whole plant physiology. The intercellular trafficking of proteins and RNAs has emerged as a novel mechanism of cell-to-cell communication in plants. As a strategy to efficient intercellular communication by cell-to-cell macromolecule trafficking, plants evolved plant cell specific symplasmic communication networks: plasmodesmata (PD) and phloem. PD are intercellular symplasmic channels connecting neighboring cytoplasms and responsible for local molecule exchange. The phloem is the sieve-tube system that allows rapid, long-distance translocation of molecules. Taken together, PD and phloem conduits have been shown to allow the exchange of proteins and RNAs, consisting of supracellular symplasmic networking. In addition to intercellular movement of viral and viroid RNAs, numerous endogenous plant RNAs including mRNAs and siRNAs were identified to move between cells, suggesting that RNA-based signaling networks may function to control many plant processes. The isolation and characterization of mRNAs from pumpkin stem phloem provided evidences that mRNA molecules circulate throughout the plant. The functional importance of long-distance movement of mRNA through the phloem was shown in *GIBBERELLIC ACID-INSENSITIVE* (*GAI*), *CmNACP* and *MOUSE EARS* (*Me*) through grafting experiments. A tomato dominant gain-of-function mutant, *Me* was used to demonstrate for long-distance movement of its mutant mRNA. *Me* mRNAs produced in mutant stock were translocated to the wild-type scions, resulting in a change in leaf morphology. Recently, trafficking of *FT* mRNA, an important component of the elusive "florigen" signal, was suggested as an essential process in flowering controlled by day-length. The day-length signal perceived in the leaf induces *FT* expression and *FT* mRNA is translocated from the leaf to the shoot apex, where downstream floral regulatory genes are activated. These findings provide insight into the presence of a novel regulatory mechanism through cell-to-cell mRNA trafficking likely used by higher plants to integrate developmental and physiological processes on a whole-plant basis.

1. INTRODUCTION

The aim of this chapter is to provide an overview of the phenomenon, mechanism and function of intercellular RNA trafficking through symplasmic pathway in higher plants. RNA molecules that are able to traffic cell-to-cell include viral and viroid RNAs, small RNAs (smRNAs) and messenger RNAs (mRNAs). In recent years, several major pieces of research have been conducted on cell-to-cell RNA trafficking in plants. Our understanding of intercellular RNA trafficking in plants grows rapidly and numerous examples underline its critical role in plant development and defense. This part introduces the general understanding of intercellular macromolecule trafficking in plants.

1.1. Pathways for Intercellular Communication

Intercellular communication in all multicellular organisms plays a critical role in the orchestration of diverse cellular processes leading to the formation and maintenance of a functional organism. A typical signaling pathway in animals and plants involves secreted signaling ligands interacting with specific receptors [1-3]. This apoplastic pathway in plants is best characterized by CLAVATA-mediated signaling, which has been shown to regulate the stem cell population in the shoot apical meristem (SAM) of *Arabidopsis* [2, 4]. In addition to this apoplastic pathway, higher plants have evolved a symplasmic signaling pathway through unique intercellular organelles, plasmodesmata (PD), which permit direct intercellular trafficking of signaling molecules. PD were initially reported to transport small molecules, such as metabolites, ions and hormones [5], but these intercellular organelles also allow selective or non-selective trafficking of macromolecules, such as proteins and RNAs [6, 7]. The communication system must present not only at a local level to coordinate events of cellular differentiation but also over long distances to coordinate developmental and physiological processes between different organs. In addition to local symplasmic networking through PD, plants evolved a long distance "information highway" called phloem. The phloem is composed of companion cells (CCs) and sieve elements (SEs), and phloem saps have been shown to include ions, nutrients, signaling molecules including hormones, regulatory proteins and RNAs [8].

1.2. Intercellular Macromolecule Trafficking

The intercellular trafficking of proteins and RNAs has emerged as a novel mechanism of cell-to-cell communication in plants [9-14]. Non-cell-autonomous proteins (NCAPs) represent all proteins that can traffic between plant cells via PD [15]. NCAPs include 1) plant viral proteins [Movement Proteins (MPs) and Coat Proteins (CPs)] 2) transcription factors [e.g. KNOTTED1 (KN1), LEAFY, DEFICIENS, GLOBOSA, CAPRICE and SHORT ROOT] and 3) phloem proteins (e.g. PP1, PP2, CmPP16, HSC70, SUT2) [16]. The first endogenous plant protein shown to move from cell to cell was a maize homeodomain protein, KN1, which is involved in the maintenance of the SAM [7]. KN1 showed abilities for its own intercellular trafficking, to increase size exclusion limit of PD and to facilitate trafficking of its own mRNA. Similar activity has already been shown in the 30kD-MP of tobacco mosaic virus (TMV) [17]. In contrast to KN1, TMV MP unspecifically binds to RNA although its native function might involve the cell-to-cell spread of the viral RNA [7, 18, 19].

The phloem sap of the angiosperms includes a variety of macromolecules, such as mRNAs, smRNAs, and proteins even though SEs lose their nuclei during development [8]. Thus, they are thought to be imported from the neighboring CCs. RNA has a prominent position in cell biology since it carries both genetic and biochemical information. Thus, RNA trafficking has broad implications in the systemic spread of infectious agents, defense responses, and the systemic regulation of gene expression controlling physiological and developmental processes in plants [20, 21]. The main part of this chapter highlights recent progress in the characterization of transport of viral and plant RNAs as components of an integrated cell-to-cell signaling network.

2. Movement of Viral RNA

The cell-to-cell movement of the plant RNA virus has been known for a long time. Thus, many studies on intercellular trafficking of RNAs focused on RNA viral infection. Plant viruses face a challenge to move from an infected cell to neighboring cells, they must traverse the plant cell wall. Specialized virus-encoded MPs control the intercellular transport of viral materials in the form of ribonucleoprotein complexes (RNP) or viral particle (virion) through PD [22, 23]. It is believed that MPs ride on an endogenous intercellular trafficking pathway. This hypothesis is supported by the discovery of CmPP16, a plant protein immunologically and functionally related to the MP of *Red clover necrotic mosaic virus*, which can traffic itself and RNA through PD [24]. This co-evolution between plant and virus makes plant viruses powerful genetic tools to explore the mechanisms controlling endogenous RNA-based signaling systems.

According to the composition of viral ribonucleoprotein complexes (VRC) that move cell-to-cell, plant viruses can be classified into three types [25]. Type I viruses including TMV, *Carmovirus* and *Hordeivirus* do not require the CP for movement. TMV encodes a specialized 30 kDa-MP and does not require CP for intercellular movement [18]. Type II viruses use CP as an (auxiliary) MP including *Cucumber mosaic virus* (CMV), *Potato virus X* (PVX), *Potyvirus*, and *Panicum mosaic virus*. CMV requires CP and MP for cell-to-cell movement. However, some mutant lacking a C-terminal portion of MP eliminates the CP requirement for movement [26]. Finally, in type III viruses including *Comoviridae*, *Alfalfa mosaic virus* , and *Brome mosaic virus*, the CP is essential because these viruses move as virion particles [27, 28]. This group encodes an MP that induces the formation of tubules in which the spherical virus particles assemble in single file through the PD, enabling to transport the virus particles to the adjacent cells [29]. In addition, some viruses such as *Cauliflower mosaic virus* (CaMV) can use more than one strategy. CaMV has characteristics typical of both type I movement of VRC and type III tubule formers. CaMV encodes a prototypical RNA binding MP that facilitates the PD opening, yet it also forms tubules through PD that contain virus particles depending on host [30]. Viruses possess their inherent flexibility to adjust the nature of the transported material to specific circumstances.

In the aspect of the mechanism of virus movement, viruses are typically divided into many classes including TMV-like, tubule formers, double or triple gene block (DGB or TGB) proteins viruses and *Potyvirus*-like [25]. Viruses using the same cell-to-cell spread strategy differ extensively in genome expression strategies and in the number and amino acid sequence of their MPs. However, recent reports suggested that smaller, segmented MPs have structural and functional similarities to specific portions of larger, single-unit MPs that encoded by TMV [31]. The notion of functional equivalence is further supported by findings that the MP of TMV can complement a movement deficiency of another, unrelated *Barley stripe mosaic virus* [32].

Although different movement strategies used in various viruses, intercellular viral trafficking share common convergent machinery limited by the plant cellular equipment available. Many VRC move along the ER membrane and actin cytoskeleton network [33]. Mechanism for recognition of PD or cell periphery by MPs has not been well demonstrated. Some MPs bind pectin methyltransferase, cell wall-associated kinase and myosin-kinesin-like protein [34]. Recently, TMV MP has been found to bind calreticulin [35]. Calreticulin is an ER lumen-located calcium-binding chaperone that can interact with integrins potentially playing a role for MP-peripheral attachment. When infected by virus, plant cells deposit callose to close PD transiently and lose callose later [36, 37]. An interaction between PVX TGB2 and an interacting protein with beta-1,3-glucase suggests a viral strategy to gate PD by callose degradation [38]. Translocation of VRC through PD pore may require chaperon activity. *Closteroviruses* encode an MP that is a HSP70 homolog and some viral MPs interact with J-domain chaperon proteins [39-41]. HSPs themselves traffic through PD and have motor activity that drives protein translocation suggesting a conserved role for HSPs in plant virus movement.

MPs unspecifically bind to RNA, thus lack specificity for recognition of the transported genome [32]. MPs are probably translated close to the genome and will bind to viral genomes in the vicinity. However, there is evidence that some viral RNAs might have a direct role in systemic movement [42]. Synonymous mutations in *Wheat streak mosaic* virus that do not affect the encoded protein affected systemic movement suggesting the requirement for specific RNA sequences in systemic infection [43].

3. Movement of Viroid RNA

Viroids, the smallest biological entity, discovered firstly by Diener [44], are single-stranded, non-coding and circular RNA molecules of which genome size ranges from 246 to 401 nucleotides (nt). Although the small size and non-translating nature of genome, viroids are able to replicate, move cell-to-cell or systemically to infect their hosts. Microinjection experiments demonstrated that *potato spindle tuber viroid* (PSTVd) RNA, moved rapidly from cell to cell between tobacco leaf mesophyll cells [45]. Since PSTVd injected into symplasmically isolated guard cells remained in the injected cells, PSTVd likely moves through PD.

Cell-to-cell movement of viroids may be mediated by a specific sequence or structural motif since PSTVd can mediate transport of a fused other cell autonomous transcript [45]. A 1.4 kb RNA containing only vector sequences was unable to move, but when PSTVd was fused to this transcript, the fusion RNA moved from cell to cell.

As viroids do not encode any protein and their movement must rely on host cellular factors, viroids provide simple models to dissect the RNA motifs that mediate cell-to-cell trafficking. Recent mutational analysis on two PSTVd mutants, PSTVdNT and PSTVdNB that differ by five nucleotides, identified an RNA motif for trafficking from bundle sheath to mesophyll cells [46]. PSTVdNB infection spreads into all types of cells in young leaves of infected tobacco, while PSTVdNT is detected only in the phloem and bundle sheath cells suggesting no trafficking into mesophyll cells. Mutational studies identified a bipartite motif that potentiates trafficking of PSTVdNB from bundle sheath into the mesophyll cells. However, this motif is not necessary for trafficking in the reverse direction or for trafficking between other cell types. Viroid infection mainly follows phloem streaming from source to sink organs. PSTVd inoculated onto the stem of tomato plants, it trafficked to sink leaves, but not source leaves. PSTVd could not enter into the SAM. In addition, in mature floral organ, PSTVd was absent in the petals, stamens, and ovary except in the sepals. This suggested that PSTVd movement does not simply follow mass flow from source to sink organs. Rather, PSTVd RNA transport into sink organs is selective and governed by plant developmental and cellular factors [47, 48].

Because the trafficking of viroids involves specific motifs, endogenous plant proteins might play a role in regulated RNA trafficking. Given that distinct RNA motifs might mediate directional trafficking between different cells, it is conceivable that at least some protein factors function in a cell-specific manner to potentiate RNA trafficking. So far a few plant proteins including the phloem lectin CsPP2, VirP1, and CmmLec17 were suggested to play a role in viroid trafficking. CsPP2 from cucumber (*Cucumis sativus* L.) binds *Hop stunt viroid* (HSVd) *in vitro* [49, 50] and *in vivo* in infected cucumber plants [51]. Both CsPP2 and HSVd can traffic long distance from rootstocks into scions in heterografts, suggesting that CsPP2 plays a role in the long-distance trafficking of viroids. A tomato protein, viroid RNA-binding protein 1 (VIRP1) interacts with the right terminal domain of PSTVd of which mutations prevented the interaction of VIRP1 and the infection of tomato [52]. These observations suggested that VIRP1 interacts with the right-terminal domain of PSTVd to facilitate trafficking [53].

A viroid most likely has evolved structural motifs that mimic endogenous plant RNA motifs so that they are recognized by cellular factors for trafficking. The dependency of PSTVd on cellular mechanisms for RNA trafficking makes them excellent tools to study the mechanisms of endogenous RNA trafficking [48, 54].

4. MOVEMENT OF SMALL RNA

4.1. Movement of Sirna and MiRNA

Post-transcriptional gene silencing (PTGS) or RNA interference (RNAi) is a sequence-specific RNA degradation pathway found in animals and plants. Systemic movement of RNA silencing has been reported in plants and *C.elegans* [55-57], where localized activation of silencing causes sequence-specific degradation of transcripts in tissues located away from the initiation zone. Silencing signal cannot enter into mature guard cells that are symplasmically isolated, suggesting its trafficking via PD [58]. However, young guard cells before their symplasmic isolation were affected by systemic signal. In addition, a strict source-to-sink direction of silencing suggested that long-distance silencing movement most probably takes place through the phloem.

In plants, PTGS was first demonstrated by grafting experiment showing the transmission of co-suppression of nitrate reductase (Nia) in tobacco [57]. Transmission was unidirectional from stock to non-silenced scion, transgene specific and locus independent, and required the presence of a transcriptionally active transgene in the target scion. Later, agro-infiltration of the transgenic *Nicotiana benthamiana* expressing GFP was found to trigger a systemic, sequence-specific loss of GFP expression [55]. From the gene-specificity of the systemic silencing, authors suggested that the signal molecule may be a nucleic acid. Supporting this idea, a single synthetic siRNA duplex, full-length sense, antisense RNA, dsRNA can trigger systemic silencing [59].

Local application of smRNA of the 20~25-nt RNA fraction from systemically silenced transgenic sunflowers also induced systemic silencing in non-silenced transgenic plants [7]. These results suggested that siRNAs themselves or intermediates induced by siRNAs could comprise silencing signals. Introduction of ectopic transgene or viral infection typically accumulates as discrete 21-nt and 24-nt species in wild type [60, 61]. However, since localized silencing was not affected in leaves co-treated with the viral P1 silencing suppressor that specifically blocks 24-nt siRNA accumulation, production of 21-nt siRNA was sufficient for localized silencing.

Micro RNA (miRNA)-mediated regulation of gene expression has been identified in both plants and animals and is currently an area of intense research activity [62]. A diverse population of smRNAs (19~25 nt) including miRNA is present in phloem sap of *Cucurbits*, *Ricinus*, *Yucca*, and lupin, suggesting a potential long-distance signaling role for small RNA [63]. However, the non-cell autonomous function of miRNAs remains controversial. In *Arabidopsis*, combined analyses employing miRNA promoter-gene fusions and miRNA sensor constructs revealed strictly overlapping transcription and activity patterns for miR171, indicating cell-autonomous functions in adult tissues [64]. It is intriguing that dsRNA-derived siRNAs had a default movement pattern of 10–15 cells in the same adult tissues where lack of miRNA movement was evidenced [60].

4.2. Local and Extensive Movement of Silencing

In plants, movement of RNA silencing could be short-distance (local) or long-distance (extensive) movement. In previous *Arabidopsis* studies, the expression of dsRNA of *RbcS* and *Sulphur* genes under the control of CC-specific *SUC2* promoter resulted in chlorosis. The chlorosis took place in around 10~15 cells outside of the veins but not expand further [60].

Cell-to-cell movement of silencing behaves differently between mature source and young sink tissues. Biolistically-induced silencing in the mature leaf of Nia transgenic tobacco plants led to localized acquired silencing forming chlorotic spots of 10~15 cells in diameter [65]. Similar observations were reported using biolistically-induced and agro-infiltration induced silencing in the mature leaf of *Nicotiana benthamiana* expressing GFP. This short-distance movement of silencing signal can take place over a limited number of cells, independent of the presence of homologous transcripts [60]. However, in sink leaves, dissemination of silencing between cells becomes extensive. Since cell-to-cell movement of silencing became extensive in mature leaves experimentally converted into sink by shading, changes in physiological status likely affect the efficiency of cell-to-cell movement in mature and developing organs.

Considering that very limited amount of trigger nucleic acid and extensive gene silencing in sink organs, initial primary signal should be amplified. The clue for signal amplification came from the discovery of 'transitive RNA silencing'. Systemic silencing not only degraded the corresponding region of target transgene, but also the non overlapping 3' region of target gene. RNA-DEPENDENT RNA POLYMERASES (RDR) now are known to be responsible for this transitivity of RNA silencing by synthesizing dsRNA triggered by primary siRNA and resulting secondary siRNA [60, 66]. In *rdr6* mutant, the spread of GFP silencing in sink tissue only affected 10~15 cells out of originally affected cells. Similar genetic approach identified that SDE3, a putative RNA helicase, also contribute to extensive silencing spread. In addition, neither the *rdr6* nor the *sde3* mutant plants with pSUC2-dsRbcS or pSUC2-dsSulphur exhibited altered phenotype of wild type. These results suggested that RDR6 was required for extensive movement of RNA silencing, but not local silencing. Himber et al [60] proposed that the extensive cell–to-cell movement of RNA silencing is likely the result of relay amplification of the 21-nt siRNA and of reiterated short-distance signaling events (figure 1) [60].

4.3. Plant Factors Required for Movement of RNA Silencing

So far, few plant endogenous proteins are reported to be directly involved in the systemic RNA silencing. Genetic mutant screening using pSUC2-dsSulphur plant to identify plant factors required for movement of siRNA silencing was carried out by Voinnet group. DCL4 was found to produce 21nt siRNAs, which probably initiated limited RNA silencing movement. And at least unidentified factors SILENCING MOVEMENT DEFICIENT (SMD1, SMD2, and SMD3) were also indicated to be required for limited RNA silencing movement [20]. These *smd* mutants accumulated the same levels of 21-nt and 24-nt siRNAs as in the pSUC2-dsSulphur parent, but they exhibited either completely lost silencing movement or extremely reduced silencing movement, indicating that the siRNAs produced in the phloem failed to spread beyond the vascular tissues in these mutants. Thus, SMD1, 2 and 3 might involve in trafficking functions of siRNAs.

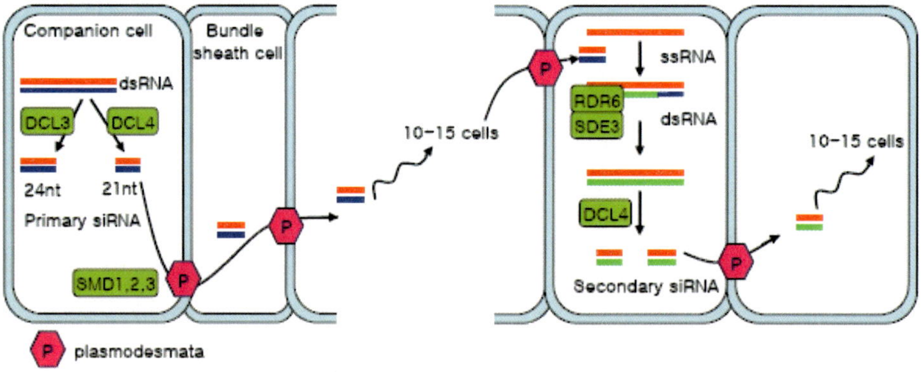

Figure 1. A re-iterative model for extensive spread of RNAi. Short-range movement can occur from original single cell to over 10~15 cells. Whereas it does not require the accumulation of 24-nt siRNAs at the site of silencing activation, the 21-nt siRNA product of DCL4 is sufficient. Movement would require SMD1, 2 and 3 and perhaps others as yet unidentified cellular factors, possibly physically associated with siRNAs. Long-range cell-to-cell movement is dependent upon RDR6 and SDE3, which use homologous transcripts as templates to produce new dsRNA. This molecule is processed into secondary siRNAs that move over 10~15 cells again. Such re-iterated short-distance signaling events would then eventually be converted into extensive, long-range movement. Redrawn from [21].

Microinjection experiments demonstrated that 25-nt siRNA do not move through PD by simple diffusion. Yoo et al. [63] purified *C. maxima* PHLOEM SMALL RNA BINDING PROTEIN1 (PSRP1), a unique component of the protein machinery probably involved in small RNA trafficking. PSRP1 can bind and facilitate cell-to-cell transport of 25-nt single-stranded small RNAs between mesophyll cells in microinjection experiment. Recent observation that the down-regulation of tomato homolog of PSRP by VIGS system resulted in the blocking of secondary VIGS by TRV-PDS suggested the involvement of PSRP in the pathways used by plants to exert systemic antiviral control (Ryu CM et al, personal communication).

It is noted that a new kind of endogenous plant siRNAs, called trans-acting (ta)siRNAs, produced by DCL4 from endogenous RDR6-dependent dsRNA, are likely functioning non-cell autonomously [67, 68]. Analysis of ta-siRNAs may provide a useful tool to dissect non-cell autonomous gene regulations by siRNAs in plants.

5. Movement of Messenger RNA

The first clue that endogenous plant mRNA might move from cell to cell came out from microinjection experiment with *KN1* mRNA [7]. *KN1* mRNA coinjected with KN1 protein into tobacco mesophyll cells showed rapid intercellular trafficking. This shined on the possibility that endogenous plant mRNA may function as non-cell autonomous signaling molecules. However, the finding on differential localization between pattern of mRNA (the L2 expression) and protein (the L1 and L2 expression) in the SAM cast a question if trafficking of *KN1* mRNA really takes place *in planta*. First *in planta* evidence came from subcellular localization study of *SUCROSE TRANSPORTER 1* (*SUT1*) using *in situ* hybridization [69]. *SUT1* mRNA synthesized CCs was detected near plasmodesmata in the SEs suggesting the movement of mRNA through PD. Numerous other transcripts, including genes controlling cell fate, the cell cycle, phytohormone response, and metabolism were subsequently identified in the pumpkin and tomato sieve tube systems [11, 70, 71]. Confirmation that these representative transcripts were phloem mobile was gained from grafting studies and RT-PCR analysis, using gene-specific primers. Indeed, this mobile population of mRNA species may be quite extensive with more than 1500 transcripts already identified from the phloem of cucurbits.

The presence of these transcripts supports the notion that RNA-based signaling networks may function to control many plant processes [72-76]. Here we provide recent major discoveries on the phloem long distance mRNA-based information network in the control of diverse development including leaf shape and flowering.

5. 1. Non-Cell Autonomous RNAs Functioning for Determination of Leaf Morphology

In angiosperms, the sieve tube system is comprised of two main cell types, the SEs and their associated CCs that are interconnected by specialized branched PD [77]. Analysis of the phloem sap collected from a number of species has demonstrated that the translocation stream contains a complex set of proteins, including a number of RNA binding proteins which have been shown to mediate the cell-to-cell transport of RNA [24, 63, 78]. Consistent with these results, a specific population of RNA molecules has been identified from the phloem translocation stream [24, 63, 69, 71]. Heterografted plants were once again employed to reveal that pumpkin stock–derived *CmNACP* transcripts were present in the cucumber scion meristem [71]. Selective translocation of RNA was confirmed by differential localization of phloem-derived transcripts in apical tissues of the cucumber scion.

Recent studies implicate a phloem long distance mRNA-based information network involved in the regulation over leaf shape. Graft transmissibility of tomato *Mouse ears* (*Me*) transcripts (*PYROPHOSPHATE-DEPENDENT PHOSPHOFRUCTOKINASE-LeT6* fusion transcripts) into wild type scions was highly correlated with the appearance of the *Me* leaf developmental phenotype [70]. Both the localization of *Me* transcripts, within the wild-type scion SAM, and the phenotype of the developing leaves reflect a property of the transcript to be translocated from stock to scion through the phloem and post-phloem transport to the SAM.

GIBBERELLIC ACID-INSENSITIVE (*GAI*) encodes a protein that belongs to the GRAS family, and functions as a negative regulator of gibberellic acid (GA) responses [79-81]. Transcripts of *GAI* in pumpkin and tomato revealed similar phloem-mobile and functional properties on leaf shape [11]. Because GA influences a wide range of plant developments [79], phloem movement of *GAI* mRNA may coordinate processes taking place in mature and young developing organs. Gain-of-function alleles, carrying mutations (gai$^{\Delta DELLA}$) with a 51bp-deletion from the conserved DELLA domain that is required for GA perception, can inhibit some components of the GA signaling pathway and result in a semidominant dwarf phenotype [81]. Long-distance signaling by the gai$^{\Delta DELLA}$ transcripts induce pronounced changes in development of newly formed leaves on the scion. In tomato system, gai$^{\Delta DELLA}$ transcripts were not detected in developing fruit, while they were detected in young leaves of the scion. Since GFP transcripts cannot traffic, gai$^{\Delta DELLA}$ mRNA translocation is a property of the transcript itself and is selective as with PSTVd RNA [46]. Both *gai*-associated developmental phenotypes and *in situ* transcript localization data, demonstrating that GAI transcripts move from the mature leaf into the meristem, proposed *GAI* mRNA as a regulator of developmental processes operating through phloem.

5.2. Non-cell Autonomous *FT* mRNA as florigen

Cell-to-cell communication between the body of the plant and the vegetative meristem through PD and phloem might be a major process to control flowering. The loading of symplasmic tracers have demonstrated that the onset of the floral transition is highly correlated with a transitory decrease in symplasmic connectivity between the apex and the terminal phloem, possibly changing the selectivity of communication [82, 83].

In many plant species, the length of the day is a major environmental cue controlling the time of flowering. In photoperiodically sensitive plants, the day-length signal is perceived by the leaves, whereas flower formation takes place in the SAM [84]. This suggested that a long-distance signal translocates from the leaves to the shoot apex resulting in vegetative-to-floral transition. This signal was coined as florigen by Chailakhyan [85-87].

In *Arabidopsis*, the zinc-finger protein, CONSTANS (CO) and the raf1 kinase inhibitor like protein (RKIP), FLOWERING LOCUS T (FT), play a central role for the photoperiodic response. Mutations in the both genes lead to late flowering under inductive long-day conditions [88]. Neither CO nor FT appears to be expressed in the SAM where floral initiation occurs. CO is expressed mainly in the vasculature of leaves [89] and induces FT in the leaf phloem [90, 91]. Grafting studies suggested that CO acts upstream of the graft-transmissible floral stimulus. While CC-specific expression of CO-GFP fusion did not reveal a capacity of CO to traffic [92], but it could induce early flowering in the co mutant background under long-day condition. Whereas CO cannot induce FT expression, when expressed from shoot apex–specific promoters [92], FT expression in both the leaf and the shoot apex can trigger floral initiation [90]. These data raise the possibility that the *FT* could be a part of the floral stimulus or control the synthesis of a mobile signal.

To test whether the *FT* transcript can move from mature leaves to the shoot apex, a heat shock-inducible transgenic system was used [93]. The application of heat-shock to a single mature leaf of p*HSP-FT* transgenic plants was sufficient to induce early flowering in the *ft* mutant background. This suggested that FT contribute to the mobile floral stimulus that moves through the phloem. In the p*HSP-FT*/ p*HSP-GUS* plants, the single-leaf heat shock treatment induced a strong transient expression of the transgenic *FT* transcript in the leaf, and an increased transgenic *FT* transcript in the shoot apex was detected 6 hours after the start of the leaf induction, but not GUS transcript. This suggested that the transgenic *FT* transcript could move from leaf to shoot apex. Interestingly, an increased level of endogenous *FT* transcript was also found during 6 to 12 hours after the leaf induction, indicating a positive auto-regulatory loop of FT expression. Further study of the activation of putative downstream targets showed that APETALA1, APETALA3, LEAFY and other genes were all induced early or late in the shoot apex after p*HSP-FT* induction in the leaf. Thus, *FT* mRNA functions as at least part of the mobile floral stimuli [93].

However, it still cannot exclude the possibility that the FT protein could also be moving and could be responsible for the floral induction. Actually, FT was detected in the phloem sap of *Brassica napus*, and the tomato FT ortholog mRNA triggers cell-autonomously systemic signals that regulate flowering. It is plausible that de novo synthesis of FT protein/mRNA in the terminal protophloem either from the imported mRNA or through positive autoregulation by imported protein, respectively, initiates the signaling relay (figure 2). How FT/*FT* mRNA moves systemically cell-to-cell and which component is essential for flowering remains to be solved.

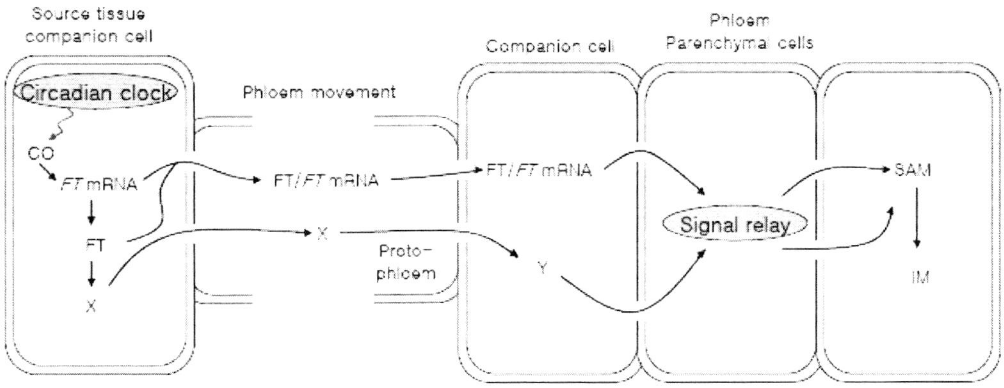

Figure 2. Florigen model on the base of phloem mobile FT trafficking in Arabidopsis. CO expression is activated under the long-day photoperiodic condition in the CCs of source leaves where it activates FT expression. Translocated FT and/or FT mRNA through phloem may serve as the florigenic stimulus. In addition, FT may exert control over floral induction through a transcriptional/post-transcriptional event within the CCs that regulates entry into the phloem of an as yet unidentified molecule (X). The positive autoregulatory loop of FT expression will serve as a signal relay system. Redrawn from [8].

6. Conclusion

The certainty that mRNA and small RNA are present, mobile in the phloem, and likely function as information macromolecules is one of the most exciting new prospects in plant biology. Clearly, RNA molecules are a powerful signal for cell-to-cell communication and local or systemic RNA trafficking through PD and phloem might have a pivotal role in plant physiology, development and plant viral defense. However, the mechanism and regulation of cell-to-cell RNA movement are largely unknown. Thus, a future challenge will be to understand how RNA trafficking occurs and how it is regulated. Functional significance of mRNA was shown in several cases among about 1500 mRNA existing in phloem sap. There is still disagreement on the non-cell autonomy of miRNA, although several miRNA were identified in phloem sap. What are the functions of non-coding RNAs found in phloem sap? Can they have any biological function? Combined proteomic/genomic, molecular genetic and biochemical approaches will be required to explore the non-cell autonomous RNA world.

ACKNOWLEDGEMENTS

Research in the authors' laboratory was supported by a Korea Research Foundation Grant (C00099), a grant of KOSEF/MOST to the Environmental Biotechnology National Core Research Center (R15-2003-012-01003-0) and National Research Laboratory (2006-02069) and a grant from BioGreen 21 Program (20050301034478), Rural Development Administration, Republic of Korea. CS and XYC are supported by scholarships from the BK21 program.

7. REFERENCES

[1] Bergey, D. R., Howe, G. A., and Ryan, C. A. (1996), Polypeptide signaling for plant defensive genes exhibits analogies to defense signaling in animals. *Proc Natl Acad Sci U S A. 93, 22*, 12053-12058.

[2] Fletcher, J. C., Brand, U., Running, M. P., Simon, R., and Meyerowitz, E. M. (1999), Signaling of cell fate decisions by CLAVATA3 in Arabidopsis shoot meristems. *Science. 283, 5409*, 1911-1914.

[3] McCarty, D. R. and Chory, J. (2000), Conservation and innovation in plant signaling pathways. *Cell. 103, 2*, 201-209.

[4] Trotochaud, A. E., Jeong, S., and Clark, S. E. (2000), CLAVATA3, a multimeric ligand for the CLAVATA1 receptor-kinase. *Science. 289, 5479*, 613-617.

[5] Robards, A. W. and Lucas, W. J. (1990), Plasmodesmata. *Annu Rev Plant Physiol Plant Mol Biol. 41*, 369-419.

[6] Wu, X., Dinneny, J., Crawford, K., Rhee, Y., Citovsky, V., Zambryski, P., and Weigel, D. (2003 Aug 15), Modes of intercellular transcription factor movement in the Arabidopsis apex. *Development. 130, 16*, 3735 - 3745.

[7] Lucas, W. J., Bouche-Pillon, S., Jackson, D. P., Nguyen, L., Baker, L., Ding, B., and Hake, S. (1995), Selective trafficking of KNOTTED1 homeodomain protein and its mRNA through plasmodesmata. *Science. 270, 5244*, 1980-1983.

[8] Lough, T. J. and Lucas, W. J. (2006), Integrative Plant Biology: Role of Phloem Long-Distance Macromolecular Trafficking. *Annu Rev Plant Biol. 57*, 203-232.

[9] Nakajima, K., Sena, G., Nawy, T., and Benfey, P. N. (2001), Intercellular movement of the putative transcription factor SHR in root patterning. *Nature. 413, 6853*, 307-311.

[10] Wada, T., Kurata, T., Tominaga, R., Koshino-Kimura, Y., Tachibana, T., Goto, K., Marks, M. D., Shimura, Y., and Okada, K. (2002), Role of a positive regulator of root hair development, CAPRICE, in Arabidopsis root epidermal cell differentiation. *Development. 129, 23*, 5409-5419.

[11] Haywood, V., Yu, T. S., Huang, N. C., and Lucas, W. J. (2005), Phloem long-distance trafficking of GIBBERELLIC ACID-INSENSITIVE RNA regulates leaf development. *Plant J. 42, 1*, 49-68.

[12] Wu, X., Weigel, D., and Wigge, P. A. (2002), Signaling in plants by intercellular RNA and protein movement. *Genes Dev. 16, 2*, 151-158.

[13] Zambryski, P. and Crawford, K. (2000), Plasmodesmata: gatekeepers for cell-to-cell transport of developmental signals in plants. *Annu Rev Cell Dev Biol. 16*, 393-421.

[14] Ding, B. (1998), Intercellular protein trafficking through plasmodesmata. *Plant Mol Biol. 38, 1-2*, 279-310.

[15] Lee, J. Y., Yoo, B. C., Rojas, M. R., Gomez-Ospina, N., Staehelin, L. A., and Lucas, W. J. (2003), Selective trafficking of non-cell-autonomous proteins mediated by NtNCAPP1. *Science. 299, 5605*, 392-396.

[16] Kim, J. Y. (2005), Regulation of short-distance transport of RNA and protein. *Curr Opin Plant Biol. 8, 1*, 45-52.

[17] Wolf, S., Deom, C. M., and Beachy, R. N. (1989), Movement Protein of Tobacco Mosaic Virus Modifies Plasmodesmatal Size Exclusion Limit. *Science. 246, 5528*, 377-379.

[18] Deom, C. M., Oliver, M. J., and Beachy, R. N. (1987), The 30-kilodalton geneproduct of tobacco mosaic-virus potentiates virus movement. *Science. 237*, 389-394.

[19] Kim, J. Y., Rim, Y., Wang, J., and Jackson, D. (2005), A novel cell-to-cell trafficking assay indicates that the KNOX homeodomain is necessary and sufficient for intercellular protein and mRNA trafficking. *Genes Dev. 19, 7*, 788-793.

[20] Dunoyer, P. and Voinnet, O. (2005), *RNA as a signalling molecule*, in A.J. Fleming, *Intercellular communication in plants*:(pp. 49-84). Cornwall, ST: Blackwell publishing

[21] Voinnet, O. (2005), Non-cell autonomous RNA silencing. *FEBS Lett. 579, 26*, 5858-5871.

[22] Lazarowitz, S. G. and Beachy, R. N. (1999), Viral movement proteins as probes for intracellular and intercellular trafficking in plants. *Plant Cell. 11, 4*, 535-548.

[23] Heinlein, M. and Epel, B. L. (2004), Macromolecular transport and signaling through plasmodesmata. *Int Rev Cytol. 235*, 93-164.

[24] Xoconostle-Cazares, B., Xiang, Y., Ruiz-Medrano, R., Wang, H. L., Monzer, J., Yoo, B. C., McFarland, K. C., Franceschi, V. R., and Lucas, W. J. (1999), Plant paralog to viral movement protein that potentiates transport of mRNA into the phloem. *Science. 283, 5398*, 94-98.

[25] Scholthof, H. B. (2005), Plant virus transport: motions of functional equivalence. *Trends Plant Sci. 10, 8*, 376-382.

[26] Nagano, H., Mise, K., Furusawa, I., and Okuno, T. (2001), Conversion in the requirement of coat protein in cell-to-cell movement mediated by the cucumber mosaic virus movement protein. *J Virol. 75, 17*, 8045-8053.

[27] Takeda, A., Kaido, M., Okuno, T., and Mise, K. (2004), The C terminus of the movement protein of Brome mosaic virus controls the requirement for coat protein in cell-to-cell movement and plays a role in long-distance movement. *J Gen Virol. 85, Pt 6*, 1751-1761.

[28] Sanchez-Navarro, J. A. and Bol, J. F. (2001), Role of the alfalfa mosaic virus movement protein and coat protein in virus transport. *Mol Plant Microbe Interact. 14, 9*, 1051-1062.

[29] Pouwels, J., van der Velden, T., Willemse, J., Borst, J. W., van Lent, J., Bisseling, T., and Wellink, J. (2004), Studies on the origin and structure of tubules made by the movement protein of Cowpea mosaic virus. *J Gen Virol. 85, Pt 12*, 3787-3796.

[30] Huang, Z., Han, Y., and Howell, S. H. (2001), Effects of movement protein mutations on the formation of tubules in plant protoplasts expressing a fusion between the green fluorescent protein and Cauliflower mosaic virus movement protein. *Mol Plant Microbe Interact. 14, 8*, 1026-1031.

[31] Morozov, S. Y. and Solovyev, A. G. (2003), Triple gene block: modular design of a multifunctional machine for plant virus movement. *J Gen Virol. 84, Pt 6*, 1351-1366.

[32] Atabekov, J. G. and Taliansky, M. E. (1990), Expression of a plant virus-coded transport function by different viral genomes. *Adv Virus Res. 38*, 201-248.

[33] Boevink, P. and Oparka, K. J. (2005), Virus-host interactions during movement processes. *Plant Physiol. 138, 4*, 1815-1821.

[34] Chen, M. H., Sheng, J., Hind, G., Handa, A. K., and Citovsky, V. (2000), Interaction between the tobacco mosaic virus movement protein and host cell pectin methylesterases is required for viral cell-to-cell movement. *Embo J. 19, 5*, 913-920.

[35] Chen, M. H., Tian, G. W., Gafni, Y., and Citovsky, V. (2005), Effects of calreticulin on viral cell-to-cell movement. *Plant Physiol. 138, 4*, 1866-1876.

[36] Iglesias, V. A. and Meins, F., Jr. (2000), Movement of plant viruses is delayed in a beta-1,3-glucanase-deficient mutant showing a reduced plasmodesmatal size exclusion limit and enhanced callose deposition. *Plant J. 21, 2*, 157-166.

[37] Bucher, G. L., Tarina, C., Heinlein, M., Di Serio, F., Meins, F., Jr., and Iglesias, V. A. (2001), Local expression of enzymatically active class I beta-1, 3-glucanase enhances symptoms of TMV infection in tobacco. *Plant J. 28, 3*, 361-369.

[38] Fridborg, I., Grainger, J., Page, A., Coleman, M., Findlay, K., and Angell, S. (2003), TIP, a novel host factor linking callose degradation with the cell-to-cell movement of Potato virus X. *Mol Plant Microbe Interact. 16, 2*, 132-140.

[39] Alzhanova, D. V., Napuli, A. J., Creamer, R., and Dolja, V. V. (2001), Cell-to-cell movement and assembly of a plant closterovirus: roles for the capsid proteins and Hsp70 homolog. *Embo J. 20, 24*, 6997-7007.

[40] Peremyslov, V. V., Hagiwara, Y., and Dolja, V. V. (1999), HSP70 homolog functions in cell-to-cell movement of a plant virus. *Proc Natl Acad Sci U S A. 96, 26*, 14771-14776.

[41] Soellick, T., Uhrig, J. F., Bucher, G. L., Kellmann, J. W., and Schreier, P. H. (2000), The movement protein NSm of tomato spotted wilt tospovirus (TSWV): RNA binding, interaction with the TSWV N protein, and identification of interacting plant proteins. *Proc Natl Acad Sci U S A. 97, 5*, 2373-2378.

[42] Kawakami, S., Watanabe, Y., and Beachy, R. N. (2004), Tobacco mosaic virus infection spreads cell to cell as intact replication complexes. *Proc Natl Acad Sci U S A. 101, 16*, 6291-6296.

[43] Choi, I. R., Horken, K. M., Stenger, D. C., and French, R. (2005), An internal RNA element in the P3 cistron of Wheat streak mosaic virus revealed by synonymous mutations that affect both movement and replication. *J Gen Virol. 86, Pt 9*, 2605-2614.

[44] Diener, T. O. and Smith, D. R. (1971), Potato spindle tuber viroid. VI. Monodisperse distribution after electrophoresis in 20 per cent polyacrylamide gels. *Virology. 46, 2*, 498-499.

[45] Ding, B., Kwon, M. O., Hammond, R., and Owens, R. (1997), Cell-to-cell movement of potato spindle tuber viroid. *Plant J. 12, 4*, 931-936.

[46] Qi, Y., Pelissier, T., Itaya, A., Hunt, E., Wassenegger, M., and Ding, B. (2004), Direct role of a viroid RNA motif in mediating directional RNA trafficking across a specific cellular boundary. *Plant Cell. 16, 7*, 1741-1752.

[47] Zhu, Y., Qi, Y., Xun, Y., Owens, R., and Ding, B. (2002), Movement of potato spindle tuber viroid reveals regulatory points of phloem-mediated RNA traffic. *Plant Physiol. 130, 1*, 138 - 146.

[48] Zhu, Y., Green, L., Woo, Y. M., Owens, R., and Ding, B. (2001), Cellular basis of potato spindle tuber viroid systemic movement. *Virology. 279, 1*, 69-77.

[49] Gomez, G. and Pallas, V. (2001), Identification of an in vitro ribonucleoprotein complex between a viroid RNA and a phloem protein from cucumber plants. *Mol Plant Microbe Interact. 14, 7*, 910-913.

[50] Owens, R. A., Blackburn, M., and Ding, B. (2001), Possible involvement of the phloem lectin in long-distance viroid movement. *Mol Plant Microbe Interact. 14, 7*, 905-909.

[51] Gomez, G. and Pallas, V. (2004), A long-distance translocatable phloem protein from cucumber forms a ribonucleoprotein complex in vivo with Hop stunt viroid RNA. *J Virol. 78, 18*, 10104-10110.

[52] Jeanmougin, F., Wurtz, J. M., Le Douarin, B., Chambon, P., and Losson, R. (1997), The bromodomain revisited. *Trends Biochem Sci. 22, 5*, 151-153.

[53] Maniataki, E., Martinez de Alba, A. E., Sagesser, R., Tabler, M., and Tsagris, M. (2003), Viroid RNA systemic spread may depend on the interaction of a 71-nucleotide bulged hairpin with the host protein VirP1. *RNA. 9, 3*, 346-354.

[54] Zhu, Y., Qi, Y., Xun, Y., Owens, R., and Ding, B. (2002), Movement of potato spindle tuber viroid reveals regulatory points of phloem-mediated RNA traffic. *Plant Physiol. 130, 1*, 138-146.

[55] Voinnet, O. and Baulcombe, D. C. (1997), Systemic signalling in gene silencing. *Nature. 389, 6651*, 553.

[56] Winston, W. M., Molodowitch, C., and Hunter, C. P. (2002), Systemic RNAi in C. elegans requires the putative transmembrane protein SID-1. *Science. 295, 5564*, 2456-2459.

[57] Palauqui, J. C., Elmayan, T., Pollien, J. M., and Vaucheret, H. (1997), Systemic acquired silencing: transgene-specific post-transcriptional silencing is transmitted by grafting from silenced stocks to non-silenced scions. *Embo J. 16, 15*, 4738-4745.

[58] Voinnet, O., Vain, P., Angell, S., and Baulcombe, D. C. (1998), Systemic spread of sequence-specific transgene RNA degradation in plants is initiated by localized introduction of ectopic promoterless DNA. *Cell. 95, 2*, 177-187.

[59] Klahre, U., Crete, P., Leuenberger, S. A., Iglesias, V. A., and Meins, F., Jr. (2002), High molecular weight RNAs and small interfering RNAs induce systemic posttranscriptional gene silencing in plants. *Proc Natl Acad Sci U S A. 99, 18*, 11981-11986.

[60] Himber, C., Dunoyer, P., Moissiard, G., Ritzenthaler, C., and Voinnet, O. (2003), Transitivity-dependent and -independent cell-to-cell movement of RNA silencing. *Embo J. 22, 17*, 4523-4533.

[61] Deleris, A., Gallego-Bartolome, J., Bao, J., Kasschau, K. D., Carrington, J. C., and Voinnet, O. (2006), Hierarchical action and inhibition of plant Dicer-like proteins in antiviral defense. *Science. 313, 5783*, 68-71.

[62] Bartel, D. P. (2004), MicroRNAs: genomics, biogenesis, mechanism, and function. *Cell. 116, 2*, 281-297.

[63] Yoo, B. C., Kragler, F., Varkonyi-Gasic, E., Haywood, V., Archer-Evans, S., Lee, Y. M., Lough, T. J., and Lucas, W. J. (2004), A systemic small RNA signaling system in plants. *Plant Cell. 16, 8*, 1979-2000.

[64] Parizotto, E. A., Dunoyer, P., Rahm, N., Himber, C., and Voinnet, O. (2004), In vivo investigation of the transcription, processing, endonucleolytic activity, and functional relevance of the spatial distribution of a plant miRNA. *Genes Dev. 18, 18*, 2237-2242.

[65] Palauqui, J. C. and Balzergue, S. (1999), Activation of systemic acquired silencing by localised introduction of DNA. *Curr Biol. 9, 2*, 59-66.

[66] Vaistij, F. E., Jones, L., and Baulcombe, D. C. (2002), Spreading of RNA targeting and DNA methylation in RNA silencing requires transcription of the target gene and a putative RNA-dependent RNA polymerase. *Plant Cell. 14, 4*, 857-867.

[67] Allen, E., Xie, Z., Gustafson, A. M., and Carrington, J. C. (2005), microRNA-directed phasing during trans-acting siRNA biogenesis in plants. *Cell. 121, 2*, 207-221.

[68] Gasciolli, V., Mallory, A. C., Bartel, D. P., and Vaucheret, H. (2005), Partially redundant functions of Arabidopsis DICER-like enzymes and a role for DCL4 in producing trans-acting siRNAs. *Curr Biol. 15, 16*, 1494-1500.

[69] Kuhn, C., Franceschi, V. R., Schulz, A., Lemoine, R., and Frommer, W. B. (1997), Macromolecular trafficking indicated by localization and turnover of sucrose transporters in enucleate sieve elements. *Science. 275, 5304*, 1298-1300.

[70] Kim, M., Canio, W., Kessler, S., and Sinha, N. (2001), Developmental changes due to long-distance movement of a homeobox fusion transcript in tomato. *Science. 293, 5528*, 287-289.

[71] Ruiz-Medrano, R., Xoconostle-Cazares, B., and Lucas, W. J. (1999), Phloem long-distance transport of CmNACP mRNA: implications for supracellular regulation in plants. *Development. 126, 20*, 4405-4419.

[72] Citovsky, V. and Zambryski, P. (2000), Systemic transport of RNA in plants. *Trends Plant Sci. 5, 2*, 52-54.

[73] Haywood, V., Kragler, F., and Lucas, W. J. (2002), Plasmodesmata: pathways for protein and ribonucleoprotein signaling. *Plant Cell. 14 Suppl*, S303-325.

[74] Jorgensen, R. A., Atkinson, R. G., Forster, R. L., and Lucas, W. J. (1998), An RNA-based information superhighway in plants. *Science. 279, 5356*, 1486-1487.

[75] Lucas, W. J., Yoo, B. C., and Kragler, F. (2001), RNA as a long-distance information macromolecule in plants. *Nat Rev Mol Cell Biol. 2, 11*, 849-857.
[76] Wu, X., Weigel, D., and Wigge, P. A. (2002), Signaling in plants by intercellular RNA and protein movement. *Genes Dev. 16, 2*, 151-158.
[77] Oparka, K. J. and Cruz, S. S. (2000), THE GREAT ESCAPE: Phloem Transport and Unloading of Macromolecules1. *Annu Rev Plant Physiol Plant Mol Biol. 51*, 323-347.
[78] Golecki, B., Schulz, A., and Thompson, G. A. (1999), Translocation of structural P proteins in the phloem. *Plant Cell. 11, 1*, 127-140.
[79] Richards, D. E., King, K. E., Ait-Ali, T., and Harberd, N. P. (2001), HOW GIBBERELLIN REGULATES PLANT GROWTH AND DEVELOPMENT: A Molecular Genetic Analysis of Gibberellin Signaling. *Annu Rev Plant Physiol Plant Mol Biol. 52*, 67-88.
[80] Pysh, L. D., Wysocka-Diller, J. W., Camilleri, C., Bouchez, D., and Benfey, P. N. (1999), The GRAS gene family in Arabidopsis: sequence characterization and basic expression analysis of the SCARECROW-LIKE genes. *Plant J. 18, 1*, 111-119.
[81] Peng, J., Carol, P., Richards, D. E., King, K. E., Cowling, R. J., Murphy, G. P., and Harberd, N. P. (1997), The Arabidopsis GAI gene defines a signaling pathway that negatively regulates gibberellin responses. *Genes Dev. 11, 23*, 3194-3205.
[82] Gisel, A., Barella, S., Hempel, F. D., and Zambryski, P. C. (1999), Temporal and spatial regulation of symplastic trafficking during development in Arabidopsis thaliana apices. *Development. 126, 9*, 1879-1889.
[83] Gisel, A., Hempel, F. D., Barella, S., and Zambryski, P. (2002), Leaf-to-shoot apex movement of symplastic tracer is restricted coincident with flowering in Arabidopsis. *Proc Natl Acad Sci U S A. 99, 3*, 1713-1717.
[84] Knott, J. E. (1934), Effect of localized photoperiod on spinach. *Proc Am Soc Hortic Sci. 31*, 152-154.
[85] Chailakhyan, M. K. (1936), New facts in support of the hormonal theory of plant development. *C R Acad Sci URSS. 13*, 79-83.
[86] Colasanti, J. and Sundaresan, V. (2000), 'Florigen' enters the molecular age: long-distance signals that cause plants to flower. *Trends Biochem Sci. 25, 5*, 236-240.
[87] Zeevaart, J. A. D. (1976), Physiology of flower formation. *Annu Rev Plant Physiol. 27*, 321-348.
[88] Koornneef, M., Hanhart, C. J., and van der Veen, J. H. (1991), A genetic and physiological analysis of late flowering mutants in Arabidopsis thaliana. *Mol Gen Genet. 229, 1*, 57-66.
[89] Yanovsky, M. J. and Kay, S. A. (2003), Living by the calendar: how plants know when to flower. *Nat Rev Mol Cell Biol. 4, 4*, 265-275.
[90] An, H., Roussot, C., Suarez-Lopez, P., Corbesier, L., Vincent, C., Pineiro, M., Hepworth, S., Mouradov, A., Justin, S., Turnbull, C., and Coupland, G. (2004), CONSTANS acts in the phloem to regulate a systemic signal that induces photoperiodic flowering of Arabidopsis. *Development. 131, 15*, 3615-3626.
[91] Takada, S. and Goto, K. (2003), Terminal flower2, an Arabidopsis homolog of heterochromatin protein1, counteracts the activation of flowering locus T by constans

in the vascular tissues of leaves to regulate flowering time. *Plant Cell. 15*, *12*, 2856-2865.

[92] Ayre, B. G. and Turgeon, R. (2004), Graft transmission of a floral stimulant derived from CONSTANS. *Plant Physiol. 135*, *4*, 2271-2278.

[93] Huang, T., Bohlenius, H., Eriksson, S., Parcy, F., and Nilsson, O. (2005), The mRNA of the Arabidopsis gene FT moves from leaf to shoot apex and induces flowering. *Science. 309*, *5741*, 1694-1696.

In: New Messenger RNA Research Communications
Editor: Lee B. Kwang, pp. 43-51

ISBN 978-1-60021-488-2

Chapter III

MicroRNAs in Nervous System

Yukio Akaneya[2]
Division of Neurophysiology, Department of Neuroscience, Osaka University Graduate School of Medicine, 2-2 Yamadaoka, Suita 565-0871 Japan

ABSTRACT

MicroRNAs (miRNAs) are an extensive newly recognized class of single-stranded non-coding RNAs that are evolutionarily conserved. They participate in the regulation of gene expression at the posttranscriptional level by using antisense complementarity to target mRNAs. Moreover, a combination of miRNAs and transcription factors can induce the inhibition of protein expression, a feature that cannot be achieved by conventional transcription factors alone. These ~22 nucleotide forms of miRNAs are generated by the processing of longer primary transcripts (pri-miRNAs) by the Dicer nuclease, which also processes double-stranded short interfering RNAs (siRNAs). Recent studies have shown that miRNAs have restricted expression and function. Hundreds of miRNAs have been cloned so far, and their roles have been revealed mainly in *Caenorhabditis elegans* and *Drosophila* and recently in mammals. The expression of miRNAs is under spatial and temporal regulation. Recently, brain-specific miRNAs have been found to be expressed at a limited stage of development and have specific neuronal functions. miRNAs are involved not only in the fate of neuronal cells and maintenance of tissue identity but also in the genetic etiology of neurological diseases. Elucidating the mechanisms of action of miRNAs will help not only in understanding their roles but also in developing therapies for neurological diseases in the near future. This review focuses on the roles of miRNAs in the nervous system with reference to neurological diseases.

[2] E-mail: *akaneya@nphys.med.osaka-u.ac.jp*, Tel: +81-6-6879-3661,Fax: +81-6-6879-3669.

Introduction

Based on early studies, RNAi was assumed to have a role in protecting cells against viruses and mobile genetic elements (Ketting et al., 1999; Waterhouse et al., 2001). However, the discovery and functional annotation of abundant miRNAs in non-coding RNAs that do not encode proteins have changed this view. miRNAs have been known to be expressed temporally and spatially (regionally) and participate in embryogenesis, development, apoptosis, and tumorigenesis (Ambros, 2004). Since the first discovery of the miRNA gene *lin-4* in *Caenorhabditis elegans* (*C. elegans*), hundreds of miRNAs have been identified in a wide range of organisms across different phylogenies, for example, viruses, plants, worms, and mammals. Further investigations have revealed that the RNAi mechanism involves the RNAi pathway that leads not only to postranscriptional gene silencing (PTGS) involving sequence-specific degradation of mRNA (Fire et al., 1998) but also to transcriptional gene silencing (TGS) and translational inhibition (Meister and Tuschl, 2004). In both animals and plants, genetic manipulations of Dicer or Argonaute proteins results in embryonic lethality. Microarray technology is instrumental in detecting miRNAs and has revealed not only the identity of many novel miRNAs but also the finding that even a single miRNA can decrease the levels of a large number of transcripts, which indicates the torelance of miRNA targets (Lim et al., 2005). In the future, an understanding of the mechanisms involved in the RNAi pathways and further identification and functional elucidation of miRNAs will be useful in disease therapy.

Recently, the discovery of novel miRNAs in many species, including humans, has been increasing dramatically (The miRNA Registry; URL: *http://microrna.sanger.ac.uk/sequences/index.shtml*). Guidelines for the annotation of new miRNAs from diverse organisms have been defined to distinguish miRNAs from other RNAs such as small interfering RNAs (Ambros et al., 2003).

Generation of miRNAs

To a large extent, the mechanisms involved in gene silencing by siRNA and miRNA share a common pathway. Although there are some differences between species, miRNAs are transcribed by RNA polymerase II into monocistronic or polycistronic primary-miRNAs (pri-miRNAs) with double-stranded hairpin structures containing a 5' 7-methyl guanosine cap and a 3' poly(A) tail (Cai et al., 2004; Lee et al., 2004) (figure 1). The pri-miRNAs are processed by the RNase III enzyme Drosha in complex with Pasha/DGCR8 (Lee et al., 2003; Denli et al., 2004) into precursor miRNAs (pre-miRNAs) containing ~70-nucleotides; the pre-miRNAs are then exported to the cytoplasm by Exportin-5 (Yi et al., 2003; Gregory et al., 2004; Lund et al., 2004). The pri-miRNAs are finally processed by Dicer, another RNase III enzyme, which recognizes its substrates by interacting with the 5' phosphate and 3' overhang on the pre-miRNAs (Ambros, 2004). This produces an ~22-nucleotide mature miRNA duplex of approximately the same length as single stranded siRNAs (Ketting et al., 1999 and 2001; Knight and Bass, 2001). This duplex is unraveled into an ~22-nucleotide mature miRNA in the cytoplasm by an uncharacterized helicase activity (Cullen, 2004).

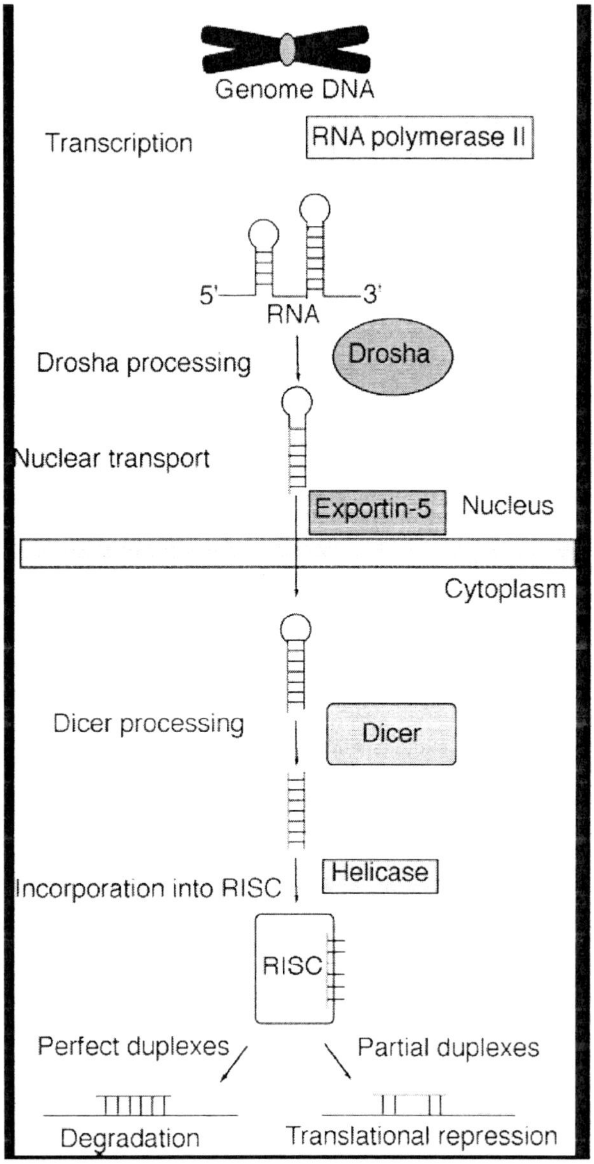

Figure 1. The basic mechanism for RNAi miRNA. RNA polymerase II transcribes genomic DNA into mRNA with hairpin-structured pri-miRNA. Initially, the nuclear RNase Drosha processes these hairpin-structured mRNA into pre-miRNAs of size approximately 70 base pairs; this is followed by the transportation of pre-miRNAs from the nucleus to cytoplasm by Exportin-5. Next, the pri-mRNA undergoes secondary processing by the Dicer RNase into almost mature miRNA of size approximately 22 base pairs; this almost mature miRNA is unwound into single-stranded RNA (miRNA) by a helicase. RISC recognizes the miRNA. The miRNA, which is guided by this complex, binds the target mRNA. If perfect miRNA-target mRNA duplex are formed, degradation occurs. However, if partial duplex are formed , transcriptional repression takes place.

One of the two strands that has looser 5' end pairing is incorporated into a micro-ribonucleoprotein complex (miRNP) known as the RNA-induced silencing complex (RISC), whereas the other strand is degraded by unknown mechanisms (Khvorova et al., 2003; Hammond, 2005). The miRNP directs the incorporated miRNA to interact with the 3' untranslated regions (UTRs) of the target mRNA by Watson-Crick base pairing. After trafficking to the GW bodies in the cytoplasm, the mRNAs that have hybridized with the miRNAs undergo degradation or translational inhibition. The higher or lower degree of sequence complementarity within the downstream nucleotides of miRNA determines either degradation or translational inhibition, respectively (Bartel, 2004), whereas the complementarity of 1–8 nucleotides of miRNA does not play a significant role in the stable interaction between miRNA and the target mRNA (Mourelatos et al., 2002; Kloosterman et al., 2004). Argonaute2 (Ago2), as the catalytic core (Slicer) in the miRNP complex, cleaves mRNA at a site near the middle of the duplex (Liu et al., 2004; Elbashir et al., 2001). As the primary mechanism, the miRNAs inhibit translation by preventing the interaction between the poly(A) tail and the 5' cap (Humphreys et al., 2005; Pillai et al., 2005). This repression by miRNAs involves accumulation of the targeted mRNAs in processing (P)-bodies (Liu et al., 2005; Sen and Blau, 2005) where the mRNAs are decapped, degraded, and stored (Sheth and Paker, 2003; Teixeira et al., 2005).

Expression and Function of Mirnas

As first shown with *lin-4* and *let-7*, miRNAs are involved in developmental fates at the embryonic stages. Recently, embryo-specific miRNAs have been shown to be involved in the lethality associated with defects in miRNA processing. For example, in the ES cells of mammals, many miRNAs are down-regulated upon cellular differentiation (Suh et al., 2004; Houbaviy et al., 2003). Analysis of the hematopoietic expression pattern in stem cells revealed the role of miR-181 in directing the fate of T cells and that of miR-223 lineages and miR-142 in regulating it (Chen et al., 2004).

Studies with *C. elegans* revealed the participation of miRNAs in neuronal symmetry in the development of the nervous system. Activation of *lys-6* (laterally symmetrical), an miRNA gene, by the transcription factor DIE-1 inhibits the expression of the gene encoding chemoreceptor GCY-5 (*gcy-5*) in the ASE left (ASEL) neuron (Johnston and Hobert, 2003; Chang et al., 2004). Conversely, the inhibition of *lys-6* results in the activation of *gcy-5* in the ASE right (ASR) neuron. Moreover, upstream of these pathways, miR-273, another miRNA, regulates DIE-1 expression.

Translation at the dendrites is necessary for synaptic plasticity such as long-term potentiation (LTP), particularly late-phase LTP, which is followed by early-phase LTP. miRNA cloning resulted in the isolation of many miRNAs from a primary culture of rat embryonic cerebral neurons, all of which cofractionated with polyribosomes (Kim et al., 2004). Moreover, it has been reported that miR-134, a brain-specific miRNA, which is localized in the synapto-dendritic compartment of hippocampal neurons down-regulates the size of dendritic spines (Schratt et al., 2006). Although many brain-specific miRNAs have been isolated using northern blots, microarray assay, and so on (Dostie et al., 2003; Kim et al., 2004; Nelson et al., 2004; Sempere et al., 2004; Krichevsky et al., 2003; Miska et al., 2004), the functions of almost all these miRNAs remain unknown.

Reference to Neurological Diseases

The most extensive research on the link between miRNAs and diseases has been conducted in the case of cancer diseases. Almost half of the characterized human miRNAs function not only to suppress but also to promote tumorigenesis (Calin et al., 2004; O'Rourke et al., 2006). Moreover, recent evidence has shown that *lin-4*, an miRNA gene, and *lin-14*, a putative transcription factor that controls the timing of larval development in *C. elegans*, also regulate the life span of the adult (Boehm and Slack, 2005). Recent studies have reported the involvement of miRNAs in some neurogenetic disorders. The putative targets can be predicted by computational algorithms used in searching neurological disorder-related genes. I will introduce several examples of miRNA-related neurological diseases that have been published recently.

Microarray manipulations using miRNAs derived from postmortem human brains have revealed abundant miRNA expression profiling (Rogaev, 2005). One of the predictions of these algorithms has revealed two candidate genes—*synapsin 1* and *FMRP1* (for fragile X mental retardation protein 1)—with putative sites for approximately 14 and 19 different miRNAs, respectively (Lewis et al., 2003). Synapsin 1, which is exclusively distributed at the synaptic terminals of axons, is known to be linked to synaptic vesicles and to be involved in synaptic plasticity. FMRP1 influences synaptic plasticity through the mechanisms of mRNA transport and translational regulation. The loss of FMRP1 causes fragile X syndrome, which is the one of the most common forms of inherited mental retardation. FMRP1 interacts with the components of Dicer and Argonaute proteins (Jin et al., 2004).

Spinal muscular atrophy (SMA), characterized by the progressive degeneration of motor neurons, is caused by deletions or loss-of-function mutations in the survival motor neuron (SMN) protein. Recently, it has been reported that miRNAs are components of miRNPs that contain Gemin3 (a DEAD-box putative RNA helicase), Gemin4, and elF2C2, a member of the Argonaute family of proteins (Mourelatos et al., 2002; Charroux et al., 1999 and 2000). Gemin3 and Gemin4 are also components of the SMN complex that functions in the assembly/restructuring of RNPs. Therefore, deletion or loss-of-function mutations of SMN in SMA may be responsible for the effects on the activity of miRNPs. Moreover, the gene locus of miR-175 has been shown to be a candidate region for neurological genetic diseases such as early-onset parkinsonism (Waisman syndrome) and X-linked mental retardation (Dostie et al., 2003).

Tourrette's syndrome is a potentially debilitating developmental neuropsychiatric disorder that is characterized by chronic vocal and motor tics with obsessive-compulsive disorder, attention-deficit hyperactivity, or substantial depression. *Slit and Trk-like 1* (*SLITRK1*) is a candidate gene for this disorder. The human miRNA hsa-miR-189 can bind the *SLITRK* mutant in which a G:U wobble base pair is replaced with an A:U Watson-Crick pairing in the has-miR-189-binding domain, resulting in a reduction in SLITRK1 expression (Abelson et al., 2005). Moreover, SLITRK1 mRNA expression overlaps with miR-189 expression in the brain regions that are closely related to this disorder, whereas SLITRK1 overexpression in the primary developing cortical neurons in cultures enhances the outgrowth of dendrites.

The DiGeorge/velocardiofacial syndrome is associated with attentional problems and executive dysfunction and is one of the highest known risk factors for schizophrenia. DGCR8 is a critical region protein for this disorder and contains two dsRNA-binding motifs. DGCR8, which can bind to Drosha, is required for the maturation of the miRNA primary transcript (Landthaler et al., 2004).

Thus, accumulating evidences have shown the relationship between miRNAs and neurogenetic diseases. Although the identification of miRNAs requires great effort, identification of its targets is even more difficult and relies on a bioinformatics approach (John et al., 2004; Kiriakidou et al., 2004; Krek et al., 2005; Lewis et al., 2003). These miRNAs can be used as biomarkers of diseases, and the target mRNAs can be the potential targets of drugs. Many miRNAs remain to be identified in non-coding RNAs, and they will provide us with genetic information on diseases.

ACKNOWLEDGMENTS

This work was supported by a Grant-in-Aid for Scientific Research from the Ministry of Education, Culture, Sports, Science and Technology in Japan.

REFERENCES

Abelson, J. F.; Kwan, K. Y.; O'Roak, B. J.; Baek, D. Y.; Stillman, A. A.; Morgan, T. M.; Mathews, C. A.; Pauls, D. L.; Rasin, M. R.; Gunel, M.; Davis, N. R.; Ercan-Sencicek, A. G.; Guez, D. H.; Spertus, J. A.; Leckman, J. F.; Dure, L. S. 4th; Kurlan, R.;, Singer, H. S.; Gilbert, D. L.; Farhi, A.; Louvi, A.; Lifton, R. P.; Sestan, N.; State, M. W. Sequence variants in SLITRK1 are associated with Tourette's syndrome. *Science 2005*, 310, 317-320.

Ambros, V.; Bartel, B.; Bartel, D. P.; Burge, C. B.; Carrington, J. C.; Chen, X.; Dreyfuss, G.; Eddy, S. R.; Griffiths-Jones, S.; Marshall, M.; Matzke M.; Ruvkun, G.; Tuschl, T. A uniform system for microRNA annotation. *RNA 2003*, 9, 277-279

Ambros, V. The functions of animal microRNAs. *Nature* 2004, 431, 350-355.

Bartel, D. P. MicroRNAs: genomics, biogenesis, mechanism, and function. *Cell* 2004, 116, 281-297.

Boehm, M.; Slack, F. A developmental timing microRNA and its target regulate life span in C. elegans. *Science 2005,* 310, 1954-1957.

Cai, X.; Hagedorn, C. H.; Cullen, B. R. Human microRNAs are processed from capped, polyadenylated transcripts that can also function as mRNAs. *RNA 2004*, 10, 1957-1966.

Calin, G. A.; Sevignani, C.; Dumitru, C. D.; Hyslop, T.; Noch, E.; Yendamuri, S.; Shimizu, M.; Rattan, S.; Bullrich, F.; Negrini, M.; Croce, CM. Human microRNA genes are frequently located at fragile sites and genomic regions involved in cancers. *Proc. Natl. Acad. Sci. USA 2004*, 101, 2999-3004.

Chang, S.; Johnston, R. J. Jr.; Frokjaer-Jensen, C.; Lockery, S.; Hobert, O. MicroRNAs act sequentially and asymmetrically to control chemosensory laterality in the nematode. *Nature 2004,* 430, 785-789.

Charroux, B.; Pellizzoni, L.; Perkinson, R. A.; Shevchenko, A.; Mann, M.; Dreyfuss, G. Gemin3: A novel DEAD box protein that interacts with SMN, the spinal muscular atrophy gene product, and is a component of gems. *J. Cell Biol.* 1999, 147, 1181-1194.

Charroux, B.; Pellizzoni, L.; Perkinson, R. A.; Yong, J.; Shevchenko, A.; Mann, M.; Dreyfuss, G. Gemin4. A novel component of the SMN complex that is found in both gems and nucleoli. *J Cell Biol.* 2000, 148, 1177-1186.

Chen, C. Z.; Li, L.; Lodish, H. F.; Bartel, D. P. MicroRNAs modulate hematopoietic lineage differentiation. *Science* 2004, 303, 83-86.

Cullen, B. R. Transcription and processing of human microRNA precursors. *Mol. Cell* 2004, 16, 861-865.

Denli, AM.; Tops, B. B.; Plasterk, R. H.; Ketting, R. F.; Hannon, G. J. Processing of primary microRNAs by the Microprocessor complex. *Nature 2004*, 432, 231-235.

Dostie, J.; Mourelatos, Z.; Yang, M.; Sharma, A.; Dreyfuss, G. Numerous microRNPs in neuronal cells containing novel microRNAs. *RNA 2003*, 9, 180-186.

Elbashir, S. M.; Lendeckel, W.; Tuschl, T. RNA interference is mediated by 21- and 22-nucleotide RNAs. *Genes Dev. 2001*, 15, 188-200.

Fire, A.; Xu, S.; Montgomery, M. K.; Kostas, S. A.; Driver, S. E.; Mello, C. C. Potent and specific genetic interference by double-stranded RNA in Caenorhabditis elegans. *Nature 1998*, 391, 806-811.

Gregory, R. I.; Yan, K. P.; Amuthan, G.; Chendrimada, T.; Doratotaj, B.; Cooch, N.; Shiekhattar, R. The Microprocessor complex mediates the genesis of microRNAs. *Nature 2004*, 432, 235-240.

Hammond, S. M. Dicing and slicing: the core machinery of the RNA interference pathway. *FEBS Lett.* 2005, 579, 5822-5829.

Houbaviy, H. B.; Murray, M. F.; Sharp, P. A. Embryonic stem cell-specific MicroRNAs. *Dev. Cell 2003*, 5, 351-358.

Humphreys, D. T.; Westman, B. J.; Martin, D. I.; Preiss, T. MicroRNAs control translation initiation by inhibiting eukaryotic initiation factor 4E/cap and poly(A) tail function. *Proc. Natl. Acad. Sci. USA* 2005, 102, 16961-1696.

Jin, P.; Zarnescu, D. C.; Ceman, S.; Nakamoto, M.; Mowrey, J.; Jongens, T. A. Nelson, D. L.; Moses, K.; Warren, S. T. Biochemical and genetic interaction between the fragile X mental retardation protein and the microRNA pathway. *Nat. Neurosci. 2004*, 7, 113-117.

John, B.; Enright, A. J.; Aravin, A.; Tuschl, T.; Sander, C.; Marks, D. S. Human MicroRNA targets. *PLoS Biol. 2004*, 2, e363.

Johnston, R. J.; Hobert, O. A microRNA controlling left/right neuronal asymmetry in Caenorhabditis elegans. *Nature 2003*, 426, 845-849.

Ketting, R. F.; Haverkamp T. H.; van Luenen, H.; Plasterk, R. H. mut-7 of *C. elegans*, required for transposon silencing and RNA interference, is a homolog of Wermer syndrome helicase and RnaseD. *Cell 1999*, 99, 133-141.

Ketting, R. F.; Fischer, S. E.; Bernstein, E.; Sijen, T.; Hannon, G. J.; Plasterk, R. H. Dicer functions in RNA interference and in synthesis of small RNA involved in developmental timing in C. elegans. *Genes Dev. 2001*, 15, 2654-2659.

Khvorova, A.; Reynolds, A.; Jayasena, S. D. Functional siRNAs and miRNAs exhibit strand bias. *Cell 2003*, 115, 209-216.

Kim, J.; Krichevsky, A.; Grad, Y.; Hayes, G. D.; Kosik, K. S.; Church, G. M.; Ruvkun, G. Identification of many microRNAs that copurify with polyribosomes in mammalian neurons. *Proc. Natl. Acad. Sci. USA* 2004, 101, 360-365.

Kiriakidou, M.; Nelson, P. T.; Kouranov, A.; Fitziev, P.; Bouyioukos, C.; Mourelatos, Z.; Hatzigeorgiou, A. A combined computational-experimental approach predicts human microRNA targets. *Genes Dev. 2*004, 18, 1165-1178.

Kloosterman, W. P.; Wienholds, E.; Ketting, R. F.; Plasterk, R. H. Substrate requirements for let-7 function in the developing zebrafish embryo. *Nucleic Acids Res.* 2004, 32, 6284-6291.

Knight, S. W.;, Bass, B. L. A role for the RNase III enzyme DCR-1 in RNA interference and germ line development in Caenorhabditis elegans. *Science 2001*, 293, 2269-2271.

Krek, A.; Grun, D.; Poy, M. N.; Wolf, R.; Rosenberg, L.; Epstein, E. J.; MacMenamin, P.; da Piedade, I.; Gunsalus, K. C.; Stoffel, M.; Rajewsky, N. Combinatorial microRNA target predictions. *Nat. Genet. 2005,* 37, 495-500.

Krichevsky, A. M.; King, K. S.; Donahue, C. P.; Khrapko, K.; Kosik, K. S. A microRNA array reveals extensive regulation of microRNAs during brain development. *RNA 2003,* 9, 1274-1281.

Landthaler, M,; Yalcin, A.; Tuschl, T. The human DiGeorge syndrome critical region gene 8 and Its D. melanogaster homolog are required for miRNA biogenesis. *Curr. Biol. 2004,* 14, 2162-2167.

Lee, Y.; Ahn, C.; Han, J.; Choi, H.; Kim, J.; Yim, J.; Lee, J.; Provost, P.; Radmark, O.; Kim, S.; Kim,; V. N. The nuclear RNase III Drosha initiates microRNA processing. *Nature 2003*, 425, 415-419.

Lee, Y.; Kim, M.; Han, J.; Yeom, K. H.; Lee, S.; Baek, S. H.; Kim, V. N. MicroRNA genes are transcribed by RNA polymerase II. *EMBO J. 2004*, 23, 4051-4060.

Lewis, B. P.; Shih, I. H.; Jones-Rhoades, M. W.; Bartel, D. P.; Burge, C. B. Prediction of mammalian microRNA targets. *Cell 2003,* 115, 787-798.

Lim, L. P.; Lau, N. C.; Garrett-Engele, P.; Grimson, A.; Schelter, J. M.; Castle, J.; Bartel, D. P.; Linsley, P. S.; Johnson, J. M. Microarray analysis shows that some microRNAs downregulate large numbers of target mRNAs. *Nature 2005*, 433, 769-773.

Liu, J.; Carmell, M. A.; Rivas, F. V.; Marsden, C. G.; Thomson, J. M.; Song, J. J.; Hammond, S. M.; Joshua-Tor, L.; Hannon, G. J. Argonaute2 is the catalytic engine of mammalian RNAi. *Science 2004,* 305, 1437-1441.

Liu, J.; Rivas, F. V.; Wohlschlegel, J.; Yates, J. R. 3rd.; Parker, R.; Hannon, G. J. A role for the P-body component GW182 in microRNA function. *Nat. Cell Biol. 2005,* 7, 1261-1266.

Lund, E.; Guttinger, S.; Calado, A.; Dahlberg, J. E.; Kutay, U. Nuclear export of microRNA precursors. *Science 2004,* 303, 95-98.

Meister, G.; Tuschl T. Mechanisms of gene silencing by double-stranded RNA. *Nature 2004*, 431, 343-349.

Miska, E. A.; Alvarez-Saavedra, E.; Townsend, M.; Yoshii, A.; Sestan, N.; Rakic, P.; Constantine-Paton, M.; Horvitz, H. R. Microarray analysis of microRNA expression in the developing mammalian brain. *Genome Biol. 2004*, 5, R68.

Mourelatos, Z.; Dostie, J.; Paushkin, S.; Sharma, A.; Charroux, B.; Abel, L.; Rappsilber, J.; Mann, M.; Dreyfuss, G. miRNPs: a novel class of ribonucleoproteins containing numerous microRNAs. *Genes Dev. 2002,* 16, 720-728.

Nelson, P. T.; Baldwin, D. A.; Kloosterman, W. P.; Kauppinen, S.; Plasterk, R. H.; Mourelatos, Z. RAKE and LNA-ISH reveal microRNA expression and localization in archival human brain. *RNA 2006*, 12, 187-191.

O'Rourke, J. R.; Swanson, M. S.; Harfe, B. D. MicroRNAs in mammalian development and tumorigenesis. Birth Defects Res. *C Embryo Today 2006*, 78, 172-179.

Pillai, R. S.; Bhattacharyya, S. N.; Artus, C. G.; Zoller, T.; Cougot, N.; Basyuk, E.; Bertrand, E.; Filipowicz, W. Inhibition of translational initiation by Let-7 MicroRNA in human cells. *Science 2005*, 309, 1573-1576.

Rogaev, E. I. Small RNAs in human brain development and disorders. *Biochemistry (Mosc)* 2005, 70, 1404-1407.

Schratt, G. M.; Tuebing, F.; Nigh, E. A.; Kane, C. G.; Sabatini, M. E.; Kiebler, M.; Greenberg, M. E. A brain-specific microRNA regulates dendritic spine development. *Nature 2006*, 439, 283-289.

Sempere, L. F.; Freemantle, S.; Pitha-Rowe, I.; Moss, E.; Dmitrovsky, E.; Ambros, V. Expression profiling of mammalian microRNAs uncovers a subset of brain-expressed microRNAs with possible roles in murine and human neuronal differentiation. *Genome Biol.* 2004, 5, R13.

Sen, G. L.; Blau, H. M. Argonaute 2/RISC resides in sites of mammalian mRNA decay known as cytoplasmic bodies. *Nat. Cell Biol. 2005*, 7, 633-636.

Sheth, U.; Parker, R. Decapping and decay of messenger RNA occur in cytoplasmic processing bodies. *Science 2003,* 300, 805-808.

Suh, M. R.; Lee, Y.; Kim, J. Y.; Kim, S. K.; Moon, S. H.; Lee, J. Y.; Cha, K. Y.; Chung, H. M.; Yoon, H. S.; Moon, S. Y.; Kim, V. N.; Kim, K. S. Human embryonic stem cells express a unique set of microRNAs. *Dev. Biol. 2004*, 270, 488-498.

Teixeira, D.; Sheth, U.; Valencia-Sanchez, M. A.; Brengues, M.; Parker, R. Processing bodies require RNA for assembly and contain nontranslating mRNAs. *RNA 2005*, 11, 371-382.

Warerhouse, P. M.; Wang, M. B.; Lough, T. Gene silencing as an adaptive defense against viruses. *Nature* 2001, 411, 834-842.

Yi, R.; Qin, Y.; Macara, I. G.; Cullen, B. R. Exportin-5 mediates the nuclear export of pre-microRNAs and short hairpin RNAs. *Genes Dev.* 2003, 17, 3011-3016.

In: New Messenger RNA Research Communications
Editor: Lee B. Kwang, pp. 53-68
ISBN 978-1-60021-488-2

Chapter IV

Problems of mRNA Quantification in the Brain Using RT-PCR

Alexander V. Kulikov*[3] *and Vladimir S. Naumenko

Institute of Cytology and Genetics, Siberian Division of the Russian Academy of Sciences, Novosibirsk, Russia

ABSTRACT

RT-PCR assay is the most sensitive technique of mRNA determination. Two main problems of mRNA quantification arise: 1) choice of standard and 2) contamination of mRNA samples with genomic DNA. The first problem is usually resolved by using known amount of artificial RNA competitors which are co-transcribed and co-amplified together with the target transcripts. However, apart from technical difficulty of generating a competitor, the differences in amplification efficacies and sizes between the target sequence and the competitor can affect quantification of the corresponding PCR products. We have shown that genomic DNA solution can be used as external standard. The genomic DNA standards are more stable, simple in preparation and titration than artificial competitor. Influence of genomic DNA contamination is usually eliminated with intron-spanning primers. However this procedure is not applicable for quantification of intronless genes transcripts. We have worked out procedure to quantify genomic DNA contamination in cDNA samples using primers specific for gene encoding tryptophan hydroxylase-1 which does not expressed in the brain. These new techniques were successfully applied to study the regulation of the 5-HT_{1A} and 5-HT_{2A} serotonin receptor mRNA levels in brain structures of mice and rats.

[3] phone (7) 383 3323101, fax (7) 383 3331278, e-mail *v_kulikov@bionet.nsc.ru.*

INTRODUCTION

Quantitative reverse transcription – polymerase chain reaction (RT-PCR) is the most sensitive and universal method for measurement of rare transcripts requiring only a minor amount of test material [Wang et al., 1989; Foley et al., 1993; Zamorano et al., 1996; Bustin, 2000]. It includes two enzymatic steps, the reverse transcription (RT) and polymerase chain reaction (PCR). At the RT step the target mRNA is transcribed into cDNA using AMV or MMLV reverse transcriptases and gene-specific, random hexamer or oligo dT primers. The synthesized cDNA molecules serve as template for PCR. At the PCR step the cDNAs are amplified using a pair of primers specific for the target cDNA and Taq DNA polymerase. Ideally the amplified sequence (amplicon) is duplicated at each cycle of PCR and the number of its copies is increased in 2^n times after n cycles. Theoretically, RT-PCR could detect as low as one copy of target transcript.

In real experimental conditions the efficacy of RT-PCR is less than the theoretically expected value [Raeymaekers, 1993]. The factors decreasing RT-PCR efficacy result from differences in the loading and quality of RNA preparation, variations of RT and PCR reaction, genomic DNA contamination, errors of sample manipulation and detection of the PCR products [Zamorano et al., 1996; Bustin, 2000].

At present fluorescent detection of PCR product is predominant. The PCR product can be detected on agarose or polyacrilamide gels by its ethidium bromide or SYBR Green fluorescence. A modern technique of fluorescent detection, named real-time PCR, can measure the PCR product after each cycle of amplification using TaqMan probe, molecular beacon or DNA-binding dye fluorescence. The real-time detection is more sensitive, informative and reliable than detection of the PCR product fluorescence on gel [Orlando et al., 1998; Bustin, 2000].

The efficacy of RT reaction is low and dependent on type and quality of used reverse transcriptase and primers. The MMLV-RNase H$^-$ reverse transcriptase in which RNase H activity has been eliminated by mutation is considered to obtain the cDNA of rare mRNAs [Zamorano et al., 1996]. Another factor significantly affecting efficacy of RT reaction is type of primers: gene-specific, oligo dT or random hexamer. The former decreases background priming, but can not ensure equal transcription for various mRNAs [Bustin, 2000]. Oligo dT primers anneal to poly A sequence on the 3'-end of mRNA and they are not mRNA-specific. However, they can not be applied for priming transcripts lack the poly A tails. Moreover, when olygo dT primers is used to prime RT reaction, the efficacy of the RT reaction can be decreased if the region to be amplified is located in the 5'-end of the transcript [Zamorano et al., 1996]. Although random hexamer primers can overestimate mRNA copy number [Zhang and Byrne, 1999; Bustin, 2000], they are not sequence-specific and thought to ensure the same rate of reverse transcription for all types of RNA. It may be expected that the cDNA mix synthesized with random hexamers conserves the same proportions of various cDNAs as in the initial mRNA mix.

The efficacy of PCR reaction defining by quality of Taq polymerase, primers, ions concentrations and anneal temperature is not constant for all cycles [Bustin, 2000]. Alterations in the efficacy at early cycles lead to dramatic change in the amount of PCR product generated [Zamorano et al., 1996; Bustin, 2000].

At last, presence of genomic DNA in RNA samples can markedly restrict accuracy of the RT-PCR quantification.

The present essay discusses the problems associating with application of standards to quantitative RT-PCR and evaluation of genomic DNA contamination.

Standards in Quantitative RT-PCR

In general, quantification is a procedure which compares evaluated variables with a set of standard measures. For example the size of object is measured by comparison with a drawing scale and its mass is compared with a set of weights. Quantification of RT-PCR is a comparison of the fluorescence of target PCR product with that of standards.

Standards utilizing for calibration of the fluorescence of RT-PCR products can be classified in endogenous internal, exogenous internal and exogenous external [Zamorano et al., 1996]. Endogenous transcripts thought to be present at constant levels throughout the mRNA samples under study serve as endogenous internal standards. The transcripts of housekeeping genes such as β-actin, GAPDH or cyclophilin are frequently used as endogenous internal standards. These standards control variations in the RT reaction and normalize the data. Known amount of artificial RNA molecules added to the mRNA samples in order to control the RT and the PCR reactions serve as exogenous internal standards or competitors. They are co-transcribed and co-amplified together with the target gene transcripts in a competitive manner. On the contrary, the RNA molecules used as exogenous external standards are transcribed and amplified in separate tubes and never mix with the target transcripts. Both exogenous internal and external standards are unable to correct differences in the RNA loading and quality in different samples.

Semi-Quantitative RT-PCR with Endogenous Internal Standards

In earlier variants of semi-quantitative RT-PCR the target transcript level was evaluated by the ratio between the fluorescence intensity of the target and housekeeping gene transcripts, such as β-actin or GAPDH (as endogenous internal standard). The commonly used approach is to co-amplify the target and the reference cDNAs in the same tube in order to decrease tube-to-tube variation [Burnet et al., 1994; Garcia-Osta et al., 2000; Yamada et al., 2002]. The coamplification of the target and standard sequences is complicated by differences in their initial concentrations, internal sequence and as follows PCR efficacies [Zamorano et al., 1996]. Since the concentration of endogenous transcripts of housekeeping gene used as endogenous internal standard is usually several time higher than the target transcript, the PCR product of the endogenous standard reaches the plateau earlier than the target sequence.

We co-amplified the transcript of 5-HT_{2A} serotonin receptors and β-actin (as endogenous internal standard) to study the effect of thyroid hormone deficit on the receptor mRNA level in rat cortex [Kulikov et al., 2002; 2004]. The RT reaction was primed with random hexamer primers which ensured the same rate of reverse transcription for the 5-HT_{2A} receptor and β-actin mRNAs. It was determined that the fluorescence intensity of PCR products increased exponentially in the range of 30-33 cycles for 5-HT_{2A} receptor and 20-24 cycles for β-actin [Kulikov et al., 2004]. So, the β-actin-specific primers were added after 9 cycles of amplification with the 5-HT_{2A}-specific primers and the mix was co-amplified for the other 23 cycles. In these conditions the fluorescence intensities of both PCR products were linearly dependent on the added cDNA amounts (figure1). The procedure was sensitive enough to show that thyroid hormone deficit significantly decreased the amount of 5-HT_{2A} transcripts without effect on the β-actin transcript level (figure 2).

The semi-quantitative design is applicable to compare the data obtained only in the same experiment and under the same conditions. Alterations in experimental conditions such as Taq polymerase quality, primers used, cycle parameters and number of cycles make the results incomparable.

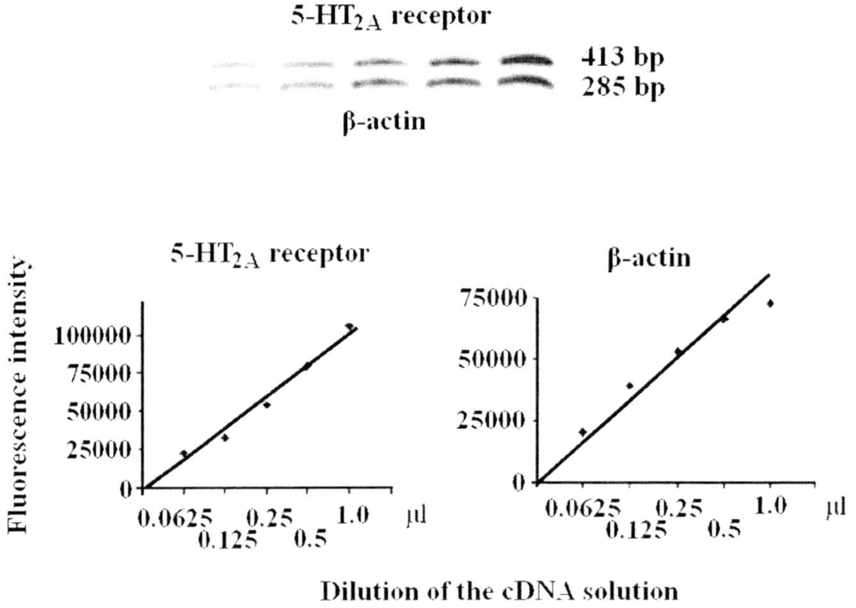

Figure 1. Fluorescence intensity of PCR product bands for 5-HT_{2A} serotonin receptor and β-actin cDNAs as a function of the amount of cDNA added to PCR mixture. The various volumes of cDNA from rat cortex was amplified with the 5-HT_{2A}-specific primers for 9 cycles, then the β-actin-specific primers were added and the mix was co-amplified for other 23 cycles. Intensity of fluorescence is presented in arbitrary units. The images on the top are negatives of the PCR product bands.

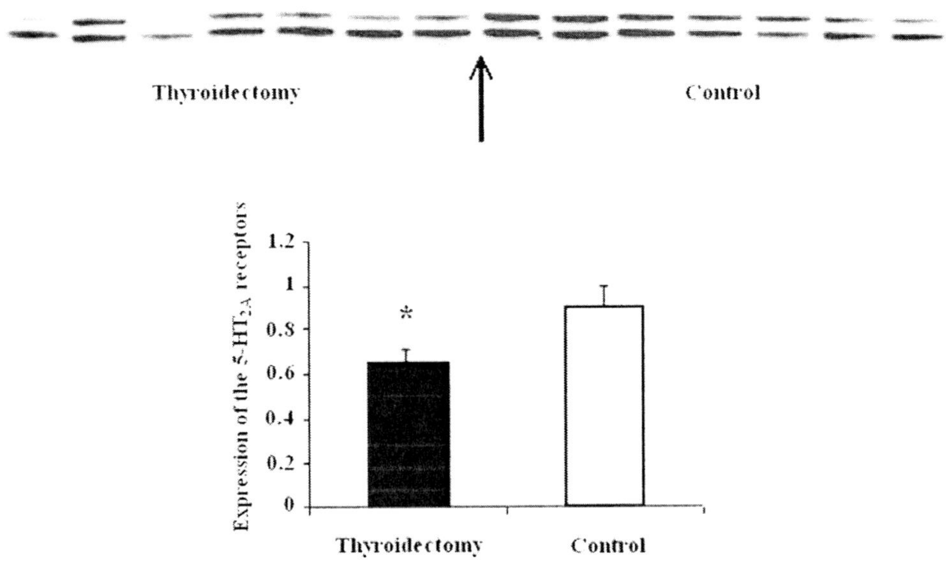

Fig.2. The expression of 5-HT_{2A} receptor in the cortex of thyroidectomized (black bar) and control sham operated (open bar) Wistar rats. The data are presented as the ratio between the 5-HT_{2A} receptor and β-actin PCR product fluorescences. The images on the top are negatives of the PCR product bands of 5-HT_{2A} receptor (upper row) and β-actin (lower row). Intensity of fluorescence is presented in arbitrary units. Data are presented as means ± S.E.M. of 8 (control) and 8 (thyroidectomized) observations. *$p<0.05$ vs control.

Competitive RT-PCR with Exogenous Internal Standards

Competitive RT-PCR with an artificial RNA sequence as exogenous internal standard thought to be able to provide an absolute quantification of the target mRNA transcripts [Wang et al., 1989; Zamorano et al., 1996]. The RNA sample containing the target transcript and the standard RNA containing the same gene-specific sequence primers as the target are reverse transcribed using either gene-specific or random hexamer primes (artificial RNA lacks the poly A tail and can not be primed with olygo dT). To distinguish the target sequence from the internal standard by electrophoresis, the latter must include a restriction site, insertion or deletion in the amplified sequence [de Kant et al., 1994; Becker-Andre and Hahlbrock, 1989]. At the same time, the target and the standard sequences should be closer as possible to ensure the same efficacy of their RT and PCR reactions. If the target and standard sequences differ more than 20%, the efficacy of their amplification should be tested carefully [Raeymaekers, 1993].

The RT mix containing the target and the standard cDNAs is amplified using gene-specific primers. The PCR products of the target and standard are separated in agarose or polyacrilamide gel and visualized by ethidium bromide fluorescence [Foley et al., 1993; Zamorano et al., 1996]. Since the standard and target sequences are virtually identical, they should be transcribed and amplified with the same efficiency and their proportion in the RNA sample should be similar to that in the PCR products. Therefore, the target transcript concentration in the RNA sample can be calibrated with the fluorescence of the standard PCR product.

The above protocol of competitive RT-PCR is thought to control the RT and PCR variations as well as the tube-to-tube errors. Sometimes, the protocol is simplified including the artificial DNA standard instead of the RNA standard. The DNA exogenous internal standard co-amplified together with the target cDNA controls the PCR reaction only [Foley et al., 1993; Zamorano et al., 1996; Carrasco et al., 1997].

Competitive RT-PCR technique was successfully applied to study the effect of serotonin transporter knockout on the expression of 5-HT_{1A} serotonin receptor gene in mouse brain structures [Li et al., 2000]. A site mutation deleting four base pairs from the sequence encoding the third intracellular loop of the mouse 5-HT_{1A} receptors was utilized as standard. To prepare 5-HT_{1A} standard RNA, the mutant sequence was transcribed with T7 RNA polymerase followed by DNase treatment and the RNA purification by phenol-chloroform extraction and ethanol precipitation. The pellet was dissolved in DEPC-treated water and the RNA concentration was determined by absorbance at 260 nm. Known amounts of the RNA standard were added to sample RNA and the mix was reverse transcribed with the gene-specific primer. Then the RT mix containing the target and standard cDNAs was amplified with the primers shared by the target and standard. To visualize PCR products, the antisense primer was end-labeled with ^{32}P-[γ]ATP. The target and standard PCR products were resolved on 8% polyacrilamide DNA sequencing gel. The gel was dried and exposed to a film. The film was analyzed densitometrically. The ratio between the density of the target and standard images was calculated and used to determine the concentration of 5-HT_{1A} mRNA in the sample. The level 5-HT_{1A} mRNA in the sample was normalized to the total RNA concentration. The authors found significant reduction of 5-HT_{1A} mRNA level in the dorsal raphe of the mutant and heterozygous mice as compared with the wild type animals [Li et al., 2000].

Last time competitive RT-PCR was criticized [Bustin, 2000]. Indeed, a large competitor concentration can markedly influence the target PCR product formation and a large series of reactions with different standard/target ratios has to be run for each unknown target concentration. There is no evidence that the amplification efficacies and the corresponding fluorescence intensities of the exogenous competitor and the target sequence PCR products are equivalent. Application of exogenous internal standard is also limited by technical difficulties of generation and cloning of mutant sequence which is virtually identical to the target, but can be distinguished by electrophoresis. At last, internal standards can be hardly applied to real-time RT-PCR.

Application of Exogenous External Standard to Real-Time RT-PCR

The progress in real-time RT-PCR increases the value of exogenous external standards. Since RT-PCRs of the target and standard transcripts are performed in different tubes, their sequences can be identical. Therefore, the efficacies of RT and PCR reactions for the target and standard can be the same. A considerable advantage of external over internal standard is simplification of calibration procedure [Bustin, 2000]. The technique of exogenous external standard application is similar to that for competitive RT-PCR except that the target and standard are transcribed and amplified in different tubes. It is thought that tube-to-tube variation resulting from sampling errors and differences in the temperature in the heat block is higher with external than with internal standard. However, numerous experiments indicate that utilizing precise pipettes and modern amplifiers can reduce tube-to-tube errors as low as 10%.

The main factor limiting universal application of internal and external exogenous standards is the technical difficulties of their generation. Cloning of the cDNA fragment in plasmid vector is frequently used for generating exogenous standard. Unfortunately, cloning is not possible in the most neurobiological and physiological laboratories. An alternative technique includes a generation of a great stock of the amplified sequence, its purification and preparation of a set of standard dilution. Another disadvantage of the cloning and amplified procedures of standard preparation is high risk of contamination of sample and equipment with amplicon or vector. It is also difficult to precise evaluate extremely low concentrations of cloned or amplified standard sequences. Moreover, exogenous standards can not normalize the target transcript concentration in the RNA samples to correct the differences in the RNA loading and quality.

Two-Standards RT-PCR with Genomic DNA as Exogenous External Standard

Theoretically, genomic DNA with known concentration can be used as an exogenous external standard, since mammalian genome contains only one copy of the most target genes. The genome sizes for human and laboratory rodents are well known. For example, mouse genome size is 2.5 x 10^9 bp which corresponds to 5 pg. So, 1 ng of mouse genomic DNA contains 200 copies of each target gene. Genomic DNA can be easily extracted from animal tissue using an overnight digestion with proteinase K followed by phenol and chloroform extraction and it concentration can be precisely measured by absorbance at 260 nm using an ordinary spectrophotometer. Genomic DNA solution is very stable and can be stored undegraded for years at -20°C. At last using genomic DNA eliminates risk of contamination.

Recently we have provided necessary experimental evidences of successful application of genomic DNA as exogenous external standard for quantitative RT-PCR [Kulikov et al., 2005; Naumenko and Kulikov, 2006]. figure 3 shows that the fluorescence of PCR products of 5-HT_{1A} receptor, β-actin and GAPDH genes are exponentially dependent on the number of cycles and linearly dependent on the amount of genomic DNA added.

The most difficult step of quantification is the selection of optimal number of cycles and standard concentration. Usually this procedure needs a large number of reactions including multiple combinations of the mentioned PCR condition. We have developed a simple routine minimizing number of manipulations needed for this selection. Firstly, the optimal cycle number is determined with 1 μl aliquot of cDNA. Secondly, the optimal concentrations of genomic DNA standard for the cycle number are determined.

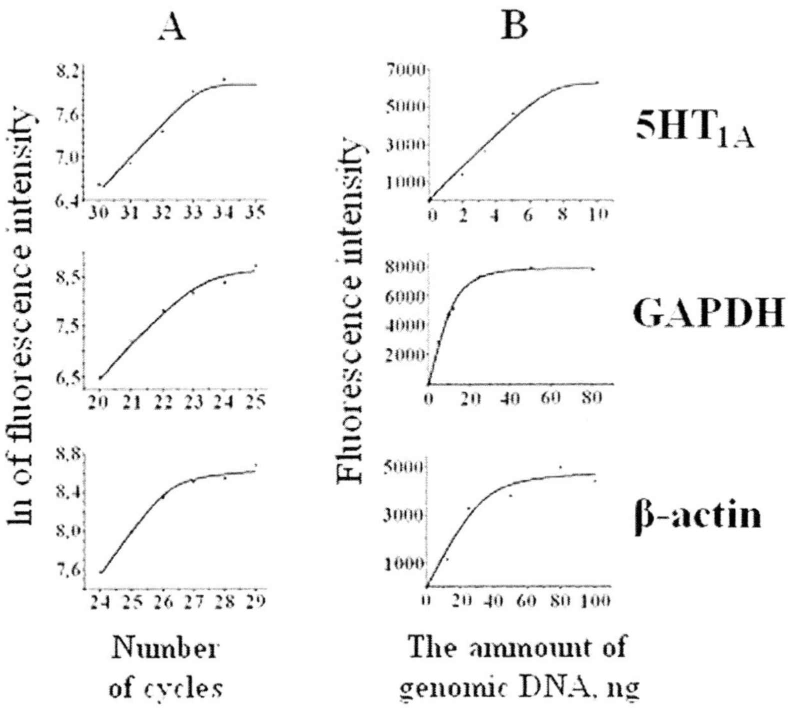

Figure 3. Fluorescent intensity of PCR product bands amplified from genomic DNA standards with the primers specific for the 5-HT_{1A} receptor, GAPDH and β-actin genes as a function of cycle number (A) and genomic DNA amount (B). Intensity of fluorescence is presented in arbitrary units (B) or their natural logarithms (A).

Genomic DNA standard can not control variability in the RT reaction and normalize the target gene expression in different RNA samples. The two-standards approach including a combination of exogenous external and endogenous internal standards was worked out to resolve these problems. Briefly, the absolute concentrations of the target and the housekeeping gene (for example β-actin) cDNAs are determined in different aliquots of the same cDNA sample using known concentrations of genomic DNA standard and corresponding primer pairs (for the target sequence and β-actin). Then the concentration of target sequence is normalized to the concentration of housekeeping gene (endogenous internal standard). This procedure allows to control both RT and PCR reactions as well as to normalize the differences in the RNA loading and quality. The data obtained with two-standards RT-PCR quantification are independent on reaction conditions, cycle number and standard concentrations and can be compared with results of other experiments.

An example of successful application of the two-standards method to study effect of chronic thyroxine treatment on the 5-HT_{2A} serotonin receptor mRNA concentration in mouse brain is shown on the figure 4. Adult mouse males of AKR strains were given T4 in drinking water for 60 days, control animals received water. The total RNA samples from the frontal cortex were reverse transcribed with random hexamers. The amounts of the receptor and β-actin cDNAs were determined in different aliquots of the same cDNA mix by PCR with corresponding primers and genomic DNA as exogenous external standard. The 5-HT_{2A}

receptor transcripts concentration was normalized to the β-actin transcripts concentration. We found a significant increase of 5-HT_{2A} transcripts in the thyroxine treated animals compared to controls.

One disadvantage of genomic DNA standard is that both primers must be placed on the same exon, since intron-spanning primers can decrease the efficacy or completely prevent amplification of the standard. Therefore, the application of such standard is limited by contamination of the RNA sample by genomic DNA.

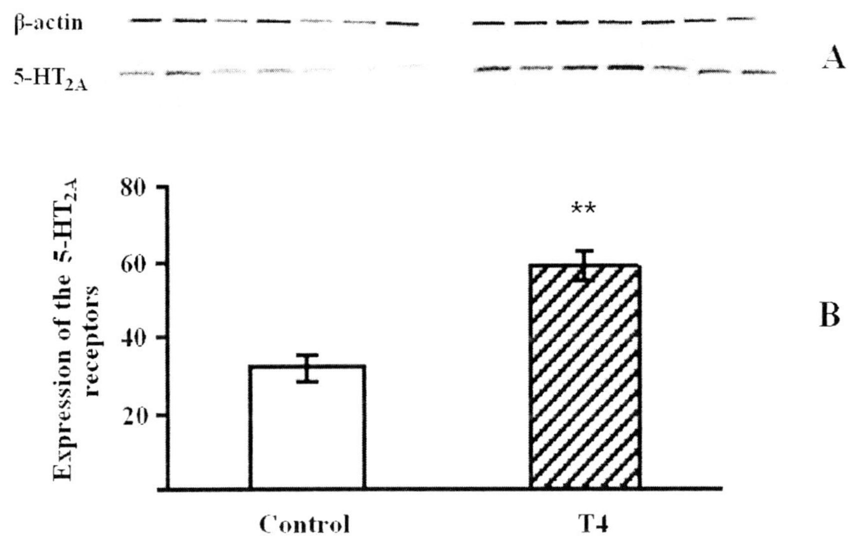

Figure 4. Effect of chronic T4 treatment on 5-HT_{2A} receptor mRNA level in the frontal cortex of AKR mice. (A) The negatives of the PCR product bands of frontal cortex samples of control and T4-treated mice. (B) The expression 5-HT_{2A} receptor presented as a number of 5-HT_{2A} receptor cDNA copies per 100 copies of β-actin cDNA. The number of receptor and β-actin copies was evaluated by calibration on the fluorescence of corresponding PCR products on the genomic DNA standards. Data are presented as means ± S.E.M. of 7 (control) and 8 (T4-treated) observations. ** $p<0.01$ vs control.

Genomic DNA Contamination

Contamination of RNA preparations by genomic DNA limits the use of RT-PCR to quantify rare transcripts. The effect of genomic DNA on the estimation of the mRNA level is most commonly eliminated by using intron-spanning primers, which specifically anneal to two (neighboring) exons separated by an intron large enough to prevent amplification from genomic DNA because of limited elongation time [Zamorano et al., 1996]. Yet this approach is inapplicable in the case of intronless genes, which include important genes encoding some serotonergic and adrenergic receptors and many other [Kobilka et al., 1987; Schofield et al., 1987; Frielle et al., 1987]. RNA specimens are usually tested for contamination with genomic DNA by PCR with primer directed to housekeeping genes, for example β-actin. Since β-actin mRNA transcripts can not be amplified, the PCR products of RNA preparation with primers specific for β-actin display its contamination with genomic DNA.

This method can not, however, determinate the genomic DNA concentration and, consequently, to estimate the RT-PCR error caused by such contamination in cDNA samples. To quantify genomic DNA contamination in cDNA one needs to use primers specific to sequence which never expressed in studied tissue. To estimate the copy number of genomic DNA in cDNA preparations from mammalian nervous tissue, we used primers specific for the fifth and sixth exons of *tph1* gene, which encoded tryptophan hydroxylase-1, the key enzyme of serotonin synthesis in peripheral tissues [Fitzpatrick, 1999]. Our choice was based on the fact that *tph1* is not expressed in the brain [Walther et al., 2003; Walther and Bader, 2003]. *Tph1* sequence includes the short (151 bp) fifth intron [Jiang et al., 2003], which makes possible to distinguish the PCR products amplified from genomic DNA (315 bp) and cDNA (164 bp) (figure 5).

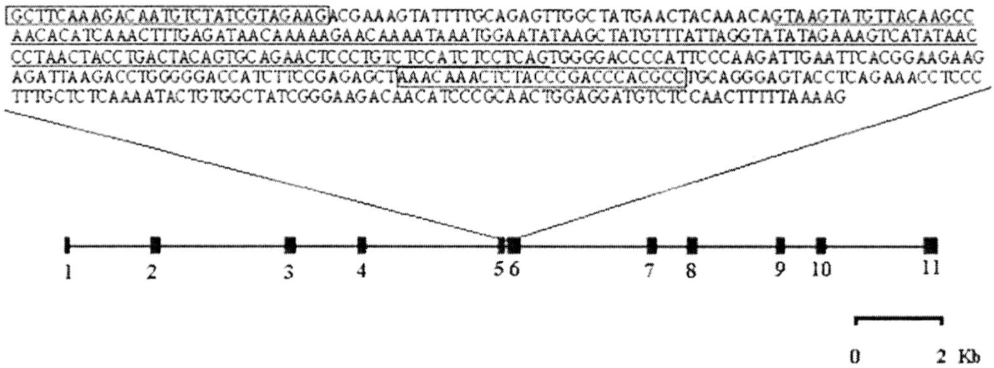

Figure 5. The mouse tryptophan hydroxylase 1 gene map. Exons marked by numbers. The nucleotide sequence of the gene fragment containing the fifth and the sixth exons separated by the fifth intron showed on the top of the picture. The sequences of forward and reversed primers are framed. The fifth intron marked by accentuation. This map based on the published sequence genomic DNA fragment encoding tryptophan hydroxylase 1 [Jiang et al., 2003].

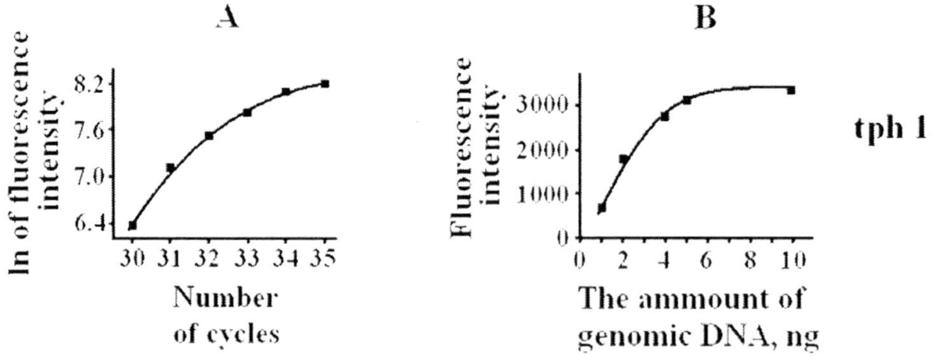

Figure 6. Fluorescence intensity of PCR products amplified from genomic DNA standards with the primers specific for the *tph1* gene as a function of cycle number (A) and genomic DNA amount (B). Intensity of fluorescence is shown in arbitrary units (B) and their natural logarithms (A).

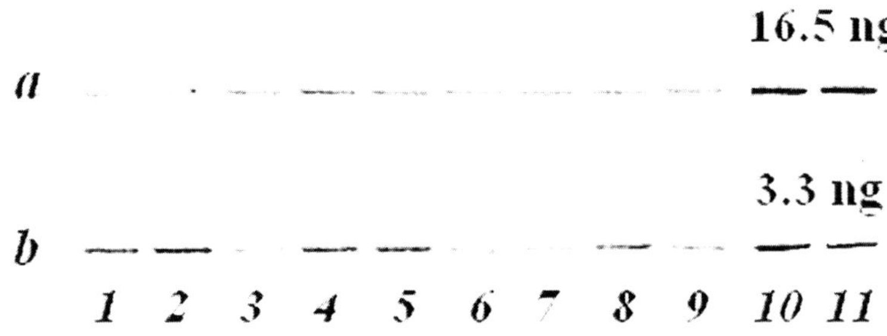

Figure. 7. Genomic DNA content in the total RNA (a) and the derivative cDNA (b) preparations. An aliquot containing 30 ng of total RNA and a 1-μl aliquot of cDNA were amplified for 22 cycles with the primers directed to the β-actin gene and for 32 cycles with the primers directed to *tph1*. The total RNA samples were isolated from the brain cortexes of C3H/He males without DNase treatment. The standards contained 16.5 (a) and 3.3 (b) ng of genomic DNA. Lines 1–9: the PCR products of total RNA and cDNA preparations obtained from different animals. Lines 10 and 11: the PCR products of the standards.

Figure 6 shows that the fluorescence of PCR products of *tph1* gene is exponentially dependent on the number of cycles and linearly dependent on the amount of genomic DNA added. We studied the copy number of genomic DNA in the samples of total RNA extracted from mouse cortex with guanidinium isothiocyanate, phenol and chlorform (figure 7) and those treated with RNase free DNase (figure 8). Genomic DNA occurred at 366.7 ± 30.4 copies per 30 ng in the total RNA isolated without RNase treatment (Fig. 7a) and at 253.3 ± 46.1 copies/μl in the resulted cDNA preparations (1 μl of cDNA corresponds to 30 ng of total RNA) (Fig. 7b). The two values did not differ significantly ($p>0.05$). So, both methods yield the same result for total RNA and derivative cDNA. The copy number of genomic DNA in the cDNA preparations obtained from total the RNA without DNase treatment is comparable with the concentration of intronless 5-HT_{1A} receptor cDNA in mouse brain (from 394 to 624 copies/μl) and, consequently, can seriously distort estimation [Naumenko and Kulikov, 2006].

After digestion with DNase, the concentration of genomic DNA in the total RNA samples quantified with the β-actin-specific primers was reduced to trace amounts (42.4 ± 17.0 copies per 30 ng) (figure 8a). The content of genomic DNA in these cDNA specimens estimated with the *tph1*-specific primers was beyond the sensitivity of our method (figure 8b). Thus, digestion of total RNA with DNase is essential for quantification of rare transcripts of intronless gene (such as 5-HT_{1A} receptor gene) in the brain.

The final procedure of two-standard quantitative RT-PCR including the control of genomic DNA contamination in RNA and the derivative cDNA preparations is shown on the figure 9.

We have successfully applied this procedure to study the effect of MAOA gene knockout on the expression of intronless 5-HT_{1A} receptor gene. The increase of 5-HT_{1A} receptor mRNA level was found in cortex and amygdala of MAO A knockout mice (Tg8 strain) compared with wild type animals of C3H/He strain [Naumenko et al., 2005; Naumenko and Kulikov, 2006]. The procedure was applied to investigate the effect of selective breeding for high and low fear-induced aggression towards man on the 5-HT_{1A} receptor mRNA level in the brain of wild Norway rats. It was found increased 5-HT_{1A} receptor mRNA level in the midbrain of the tame rats compared to the animals bred for high aggression [Popova et al., 2005].

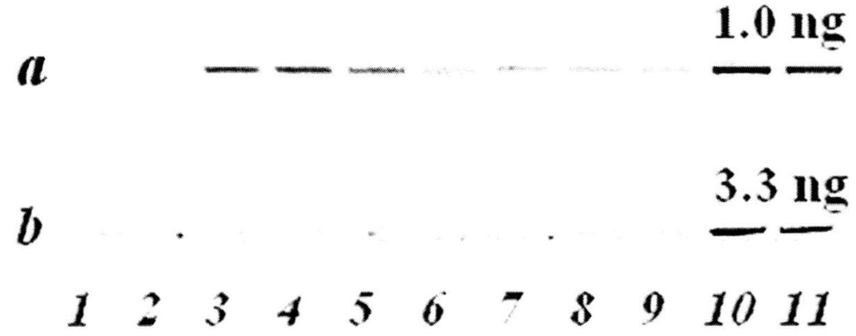

Figure 8. Genomic DNA content in the DNase-treated total RNA (a) and the derivative cDNA (b) preparations. An aliquot containing 30 ng of total RNA and a 1-μl aliquot of cDNA were amplified for 32 cycles with the primers directed to the β-actin gene (RNA) or *tph1* (cDNA). The total RNA samples were isolated from the brain cortex of C3H/H3 males and digested with DNase. The standards contained 1 (a) and 3.3 (b) ng of genomic DNA. Lanes 1–9: the PCR products of total RNA and cDNA preparations obtained from different animals. Lines 10 and 11: the PCR products of the standards.

Conclusion

Although the progress in PCR instrumentation and chemistries results in substantial decrease of tube-to-tube error, RT and PCR variations, the role of standard in modern quantitative RT-PCR is still big enough. Since exogenous external standards are the most convenient for quantitative real-time RT-PCR, their role can be predicted to be increased in future. Universal utilization of artificial sequences as exogenous external standards is limited by the difficulties of their preparation and titration. An alternative approach is to use genomic DNA as exogenous external standard for controlling the PCR reaction. Genomic DNA standards have technical advantages, namely high stability, simple preparation and titration, however they can not control the RT reaction. We have provided evidences that both RT and PCR reactions can be successfully controlled with a combination of genomic DNA as exogenous external standard and transcripts of housekeeping genes as endogenous internal standard.

Contamination of RNA and the derived cDNA preparations with genomic DNA presents considerable problem for the quantification of intronless gene transcripts. We have found a relatively high concentration of genomic DNA in the RNA samples prepared with guanidinium isothiocyanate-phenol-chloroform extraction. Even treatment of the RNA samples with DNase is unable to eliminate genomic DNA completely and this contamination is conserved in the derivative cDNA preparations.

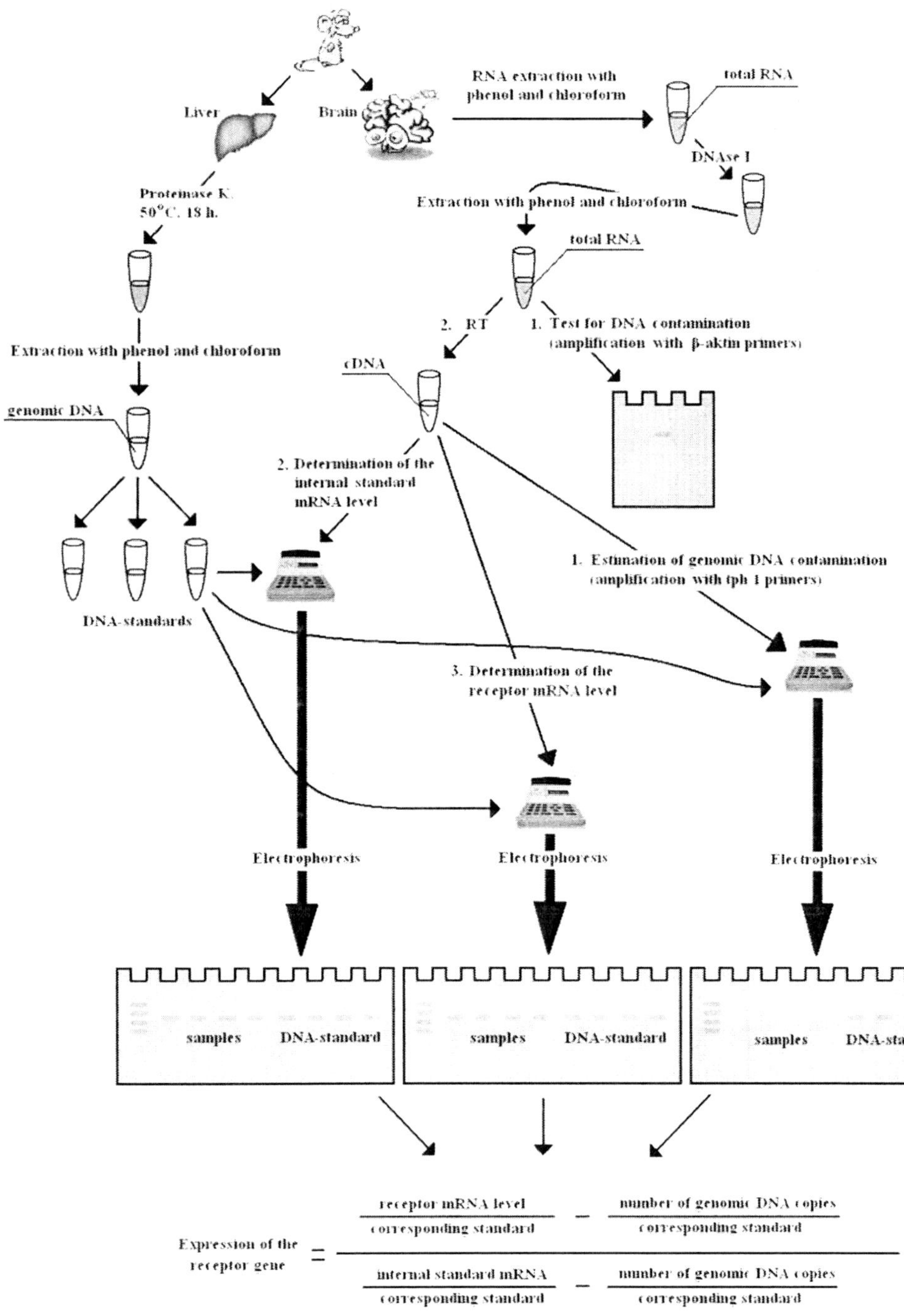

Figure 9. Scheme of two-standard RT-PCR technique including correction to genomic DNA contamination.

One problem of quantitative RT-PCR is normalization of the target transcript level which serves to correct data to possible variation of the RNA loading and quality. Many authors used endogenous transcripts of β-actin or GAPDH for calibraiting the target transcript concentration. However, this method fails when the expression of β-actin or GAPDH is modified by genotype, age, tissue and experimental treatment [Mansur et al., 1993; Spanakis, 1993; Calvo et al., 1997]. Some authors advice in these cases to use total RNA concentration as normalizing factor [Bustin, 2000]. At the same time, in physiological experiment the alterations in β-actin or GAPDH transcription induced by some treatment may be considered as a measure of nonspecific effect. Therefore, the changes of the target transcript level against background of housekeeping gene transcript concentration may be informative for detection of the specific effect of the treatment on the target gene expression.

ACKNOWLEDGEMENTS

The work was partially supported by grants of Siberian Branch of Russian Academy of Sciences (Integration project N°5) and RFBR (grant N°07-04-00209).

REFERENCES:

Becker-Andre, M. and Hahlbrock, K. (1989). Absolute mRNA quantification using the polymerase chain reaction (PCR). A novel approach by a PCR aided transcript titration assay (PATTY). *Nucleic Acids Res.*, V.17(22). P.9437-9446.

Burnet, P.W.J., Eastwood, S.L. and Harrison, P.J. (1994). Detection and quantitation of 5-HT_{1A} and 5-HT_{2A} receptor mRNAs in human hippocampus using a reverse transcription-polymerase chain reaction (RT-PCR) technique and their correlation with binding site densities and age. *Neurosci. Lett.,* V.178. P.85-89.

Bustin, S. (2000). Absolute quantification of mRNA using real-time reverse transcription polymerase chain reaction assays. *J. Mol. Endocrinol.,* V.25. P.169-193.

Calvo, E.L., Boucher, C., Coulombe, Z. and Morisset, J. (1997). Pancreatic GAPDH gene expression during ontogeny and acute pancreatitis induced by caerulein. Biochem. Biophys. *Res. Commun.*, V.235(3). P.636-640.

Carrasco, C., Esteban, O. and Domingo, A. (1997). *In Vitro* generation of competitor DNA segments for the quantitation of any nucleic acid known sequence by competitive polymerase chain reaction. *Anal. Biochem.*, V.244. P.406-407.

de Kant, E., Rochlitz, C.F. and Herrmann, R. (1994). Gene expression analysis by a competitive and differential PCR with antisense competitors. *Biotechniques*, V.17(5). P.934-942.

Fitzpatrick, P.F. (1999). Tetrahydopterin-dependent amino acid hydroxylases. *Ann. Rev. Biochem.*, V.68. P.355-381.

Foley, K.P., Leonard, M.W. and Engel, J.D. (1993). Quantitation of RNA using the polymerase chain reaction. *Trends Genet.*, V.9. P.380-385.

Frielle, T., Collins, S., Daniel, K.W., Caron, M.G., Lefkowitz, R.J. and Kobilka, B.K. (1987). Cloning of the cDNA for the human beta 1-adrenergic receptor. *Proc. Natl. Acad. Sci.* USA, V.84(22). P.7920-7924.

Garcia-Osta, A., Frechilla, D. and Del Rio, J. (2000). Effect of p-chloroamphetamine on 5-HT_{1A} and 5-HT_7 serotonin receptor expression in rat brain. *J. Neurochem.*, V.74. P.1790-1797.

Jiang, X., Song, L., Ni, Y. and Roe, B. Mus musculus clone rp23-28a7 map 7 strain C57BL/6J. EMBL/GenBank/DDBJ/AC090122.26. 2003. Available from: *http://www.ebi.ac.uk/fasta33/nucleotide.html.*

Kobilka, B.K., Frielle, T., Collins, S., Yang-Feng, T., Kobilka, T.S., Francke, U., Lefkowitz, R.J. and Caron, M.G. (1987). An intronless gene encoding a potential member of the family of receptors coupled to guanine nucleotide regulatory proteins. *Nature,* V.329(6134). P.75-79.

Kulikov, A.V., Maksyutova, A.V., Ivanova, E.A., Khvorostov, I.B. and Popova, N.K. (2002). The effect of thyroidectomy on the expression of the mRNA of 5-HT_{2A} serotonin receptors in the rat frontal cortex. Dokl. Biochem. *Biophys.*, V.383. P.116-118.

Kulikov, A.V., Naumenko, V.S., Voronova, I.P., Tikhonova, M.A. and Popova, N.K. (2005). Quantitative RT-PCR assay of 5-HT_{1A} and 5-HT_{2A} serotonin receptor mRNAs using genomic DNA as an external standard. *J. Neurosci. Methods,* V.141(1). P.92-101.

Kulikov, A.V., Tikhonova, M.A., Voronova, I.P. and Popova, N.K. (2004). The application of semi-quantitative RT-PCR assay for estimation of serotonin 5-HT_{2A}-receptors mRNA expression in the brain. *Neirokhimiya,* 2004. V.21(12). P.76-79. (in Russian).

Li, Q., Wichems, C., Heils, A., Lesch, K-P. and Murphy, D.L. (2000). Reduction in the density and expression, but not G-protein coupling, of serotonin receptors (5-HT_{1A}) in 5-HT transporter knock-out mice: gender and brain region differences. *J. Neurosci.*, V.20. P.7888-7895.

Mansur, N.R., Meyer-Siegler, K., Wurzer, J.C. and Sirover, M.A. (1993). Cell cycle regulation of the glyceraldehyde-3-phosphate dehydrogenase/uracil DNA glycosylase gene in normal human cells. *Nucleic Acids Res.*, V.21(4). P.993-998.

Naumenko, V.S. and Kulikov, A.V. (2006). Quantitative assay of 5-HT_{1A} receptor gene expression in the brain. *Mol. Biol.*, V.40. P.37-44.

Naumenko, V.S., Ivanova, E.A., Kulikov, A.V. and Popova, N.K. (2005). Effect of monoamine oxidease A knockout on the expression of 5-HT_{1A} receptors. *Dokl. Biol. Sci.*, V.402. P.205-207.

Orlando, C., Pinzani, P. and Pazzagli, M. (1998). Developments in quantitative PCR. *Clin. Chem. Lab. Med.*, V.36(5). P.255-269.

Popova, N.K., Naumenko, V.S., Plyusnina, I.Z. and Kulikov, A.V. (2005). Reduction in 5-HT(1A) receptor density, 5-HT(1A) mRNA expression, and functional correlates for 5-HT(1A) receptors in genetically defined aggressive rats. *J. Neurosci. Res.*, V.80(2). P.286-292.

Raeymaekers, L. (1993). Quantitative PCR: theoretical considerations with practical implications. *Anal. Biochem.*, V.214. P.582-585.

Schofield, P.R., Rhee, L.M. and Peralta, E.G. (1987). Primary structure of the human beta-adrenergic receptor gene. *Nucleic Acids Res.*, V.15(8). P.3636.

Spanakis, E. (1993). Problems related to the interpretation of autoradiographic data on gene expression using common constitutive transcripts as controls. *Nucleic Acids Res.*, V.21(16). P.3809-3819.

Walther, D. and Bader, M. (2003). A unique central tryptophan hydroxylase isoform. *Biochem. Pharmacol.*, V.66. P.1673-1680.

Walther, D.J., Peter, J.-U., Bashammakh, S., Hortnagl, H., Voits, M., Fink, H. and Bader, M. (2003). Synthesis of serotonin by a second thryptophan hydroxylase isoform. *Science,* V.299. P.76.

Wang, A.M., Doyle, M.V. and Mark, D.F. (1989). Quantitation of mRNA by the polymerase chain reaction. *Proc. Natl. Acad. Sci.* USA., V.86(24). P.9717-9721.

Yamada, K., Wada, E., Yamano, M., Sun, Y-J., Ohara-Imaizumi, M., Nagamatsu, S. and Wada, K. (2002). Decreased marble burying behavior in female mice lacking neuromedin-B receptor (NMB-R) implies the involvement of NMB/NMB-R in 5-HT neuron function. *Brain Res.*, V.942. P.71-78.

Zamorano, P.L., Mahesh, V.B. and Brann, D.W. (1996). Quantitative RT-PCR for neuroendocrine studies. Neuroendocrinology, V.63. P.397-407.

Zhang, J. and Byrne, C.D. (1999). Differential priming of RNA templates during cDNA synthesis markedly affects both accuracy and reproducibility of quantitative competitive reverse-transcriptase PCR. *Biochem. J., V.*337(2). P.231-241.

In: New Messenger RNA Research Communications
Editor: Lee B. Kwang, pp. 69-86
ISBN 978-1-60021-488-2

Chapter V

An Insight of Subcellular Localization of Pituitary Hormone MRNA: Observation Under Electron Microscopy and Confocal Laser Scanning Microscopy

Akira Matsuno[4], Johbu Itoh[2], Hirotoshi Utsunomiya[3], Susumu Takekoshi[4] and R. Yoshiyuki Osamura[4]

[1] Department of Neurosurgery, Teikyo University Chiba Medical Center, 3426-3 Anesaki, Ichihara City, Chiba 299-0111, Japan

[2] Teaching and Research Support Center, Tokai University School of Medicine, Boseidai, Isehara City, Kanagawa 259-1100, Japan

[3] Department of Pathology, Wakayama Medical College, 811-1 Kimiidera, Wakayama City, Wakayama 640-0012, Japan

[4] Department of Pathology, Tokai University School of Medicine, Boseidai, Isehara City, Kanagawa 259-1100, Japan

Abstract

Non-radioisotopic electron microscopic (EM) in situ hybridization (ISH) (EM-ISH) with biotinylated oligonucleotide probes is utilized for the ultrastructural visualization of pituitary hormone mRNA in rat pituitary cells. This EM-ISH is an important tool for clarifying the intracellular localization of mRNA and the exact site of specific hormone synthesis on the rough endoplasmic reticulum. The simultaneous visualization of mRNA and encoded protein in the same cell using preembedding EM-ISH and subsequent postembedding immunoreaction with protein A colloidal gold complex can provide an

[4] Correspondence: Akira Matsuno, M.D., Ph.D. Department of Neurosurgery, Teikyo University Chiba Medical Center .3426-3 Anesaki, Ichihara City, Chiba 299-0111, Japan Tel: 81-436-62-1211 ,Fax:81-436-62-1357.

important clue for elucidating the intracellular correlation of mRNA translation and secretion of translated protein. This combined ISH and immunohistochemistry (IHC) under electron microscopy (EM-ISHandIHC) has sufficient ultrastructural resolution, and provides two-dimensional images of subcellular localization of pituitary hormone and its mRNA in a pituitary cell. The advantages of semiconductor nanocrystals (Quantum dots, Qdots) and confocal laser scanning microscopy (CLSM) enable us to obtain three-dimensional images of subcellular localization of pituitary hormone and its mRNA. Both EM-ISHandIHC and ISH and IHC using Qdots and CLSM are useful for understanding the relationship between protein and mRNA simultaneously in two or three dimensions.

Key Words: pituitary hormone, mRNA, intracellular localization, rough endoplasmic reticulum.

INTRODUCTION

Observation of intracellular expression of mRNA and subcellular organelles is essential for the study of spatial relationship of mRNA and encoded protein. Electron microscopic (EM) *in situ* hybridization (ISH) (EM-ISH) is utilized for examining the intracellular distribution and role of mRNA in protein synthesis. This EM-ISH study has been carried out in the three different approaches: preembedding method [1-8], non-embedding method using ultrathin frozen sections [5,9,10], and postembedding method [5,6,8,11-13]. In recent years, we have developed non-radioisotopic EM-ISH method using biotinylated synthesized oligonucleotide probes for the purpose of the ultrastructural visualization of growth hormone (GH) and prolactin (PRL) mRNAs in rat pituitary cells, and have applied this method to pathophysiological studies of pituitary cells [6-8]. In addition, we have utilized a combined EM-ISH and immunohistochemistry (IHC) (EM-ISHandIHC), in order to investigate simultaneously the intracellular localization of pituitary hormone and its mRNA [14-21]. The EM-ISHandIHC method is very useful for the study of the spatial relationship of mRNA and the encoded protein. It has high resolution for ultrastructure, and can provide two-dimensional images of the intracellular localization of mRNA and protein.

Meanwhile, confocal laser scanning microscopy (CLSM) enables us to observe subcellular organelles, mRNA and protein, three-dimensionally, in routinely processed light microscopic specimens [22-33]. Moreover, recently developed semiconductor nanocrystals (Quantum dots, Qdots) enable us to obtain multicolor images of molecules due to a narrow emission peak that can be excited via a single wavelength of light [34,35]. In biological research, Qdots have been used for the detection of the signals of IHC and fluorescence in situ hybridization (FISH) [36-41]. Recently, we have successfully applied the above-mentioned advantages of Qdots and CLSM to three-dimensional imagings of the intracellular localization of mRNA and protein [42,43].

In this paper, we compare between the two methods: two-dimensional imagings of simultaneous visualization of rat GH and its mRNA using electron microscopy, and three-dimensional imagings using CLSM and Qdots.

TISSUE PREPARATION

Male and female Wistar-Imamichi rats (8 weeks from birth, body weight ranging from 240 to 330 g, purchased from Charles River Japan Inc., Yokohama, Japan) were studied. Male rats were used to study the intracellular localization of GH mRNA and GH protein. Female rats were used to study the intracellular localization of PRL mRNA and PRL protein. The pituitary glands were removed, and the anterior lobes were immediately fixed overnight at 4°C in 4% paraformaldehyde dissolved in 0.01 M phosphate buffered saline, pH 7.4 (PBS). After immersion in graded concentrations of sucrose dissolved in PBS at 4°C (10% for 1 h, 15% for 2 h, 20% for 4 h), the tissues were embedded in Optimal Cutting Temperature (OCT) compound (Tissue-Tek; Miles Laboratories Inc., Elkhart, Ind., USA).

Ribonuclease-free solutions treated with 0.02% diethylpyrocarbonate (DEPC) were used routinely, and gloves were used when handling all the tissue specimens and glass slides.

BIOTINYLATION OF SYNTHESIZED OLIGONUCLEOTIDE PROBES FOR ISH

The sequence of the oligonucleotide probe for rat GH mRNA is 5'-dATC GCT GCG CAT GTT GGC GTC, and the sequence of the oligonucleotide probe for rat PRL mRNA is 5'-dGGC TTG CTC CTT GTC TTC AGG [44]. The antisense, sense, and scramble oligonucleotide probes were synthesized with a DNA synthesizer (Applied Biosystems model 392; Applied Biosystems, Foster City, CA, USA) and biotinylated by 3'-end labeling method using ENZO's terminal labeling kit (ENZO Diagnostics Inc., Farmingdale, NY, USA), according to the manufacturer's protocol. The specificities of the biotinylated probes for both hormone mRNAs were confirmed by Northern blot hybridization, using total RNA extracted from normal male Wistar-Imamichi rat pituitary glands [6,7].

EM-ISH: PREEMBEDDING METHOD

Six μm thick tissue specimens were mounted on 3-aminopropylmethoxysilane-coated slides. After air drying for 1 h, tissue sections were washed with PBS for 15 min. Subsequently they were treated with 0.1 μg/ml proteinase K at 37°C for 30 min, followed by treatment for 10 min with 0.25% acetic anhydride in 0.1 M triethanolamine. The slides were washed in 2x sodium chloride sodium citrate (SSC) at room temperature for 3 min and then prehybridized at 37°C for 30 min. The prehybridization solution consists of 10% dextran sulfate, 3xSSC, 1x Denhardt's solution (0.02% Ficoll/ 0.02% bovine serum albumin (BSA)/ 0.02% polyvinylpyrrolidone), 100 μg/ml salmon sperm DNA, 125 μg/ml yeast tRNA, 10 μg/ml polyadenylic-cytidylic acid, 1 mg/ml sodium pyrophosphate pH 7.4 and 50% formamide. The biotinylated probe with the concentration of 0.1 ng/μl was diluted with this solution and hybridization was carried out at 37°C overnight. After hybridization, the slides were washed at room temperature with 2xSSC, 1xSSC and then 0.5xSSC for 15 min, respectively. The hybridization signals were detected with streptavidin-biotin-horseradish peroxidase (ABC-HRP), using Vectastain's ABC kit (Vector Laboratories Inc., Burlingame, CA, USA), and thereafter developed with diaminobenzidine (DAB) and H_2O_2. After osmification and dehydration with ethanol, tissue sections were embedded in Epon resin, which was polymerized at 60°C for 2 days. Ultrathin sections were inspected under electron microscopy. The control experiments carried out were hybridization with probes of sense or scramble sequence and without probes.

Hybridization signals for rat GH mRNA were demonstrated under light microscopy using ABC-HRP. Light microscopic ISH study with a sense probe showed negative signals. EM-ISH with an antisense probe for rat GH mRNA revealed diffuse localization of GH mRNA on the polysomes of the entire rough endoplasmic reticulum (RER) (figure.1). EM-ISH with a sense probe for GH mRNA showed negative signals (figure.2).

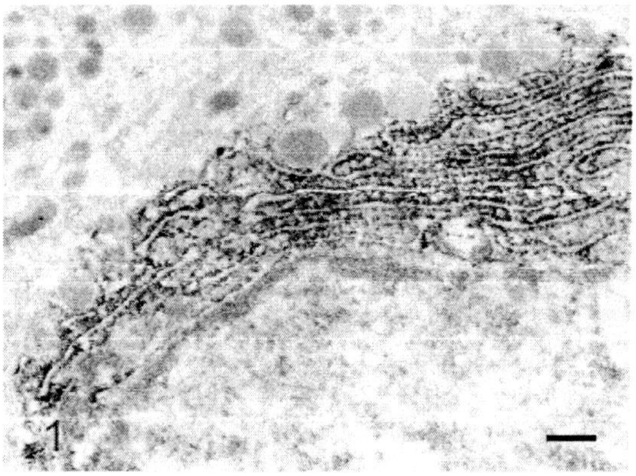

Figure 1. Photograph of preembedding EM-ISH for GH mRNA. GH mRNA is localized diffusely on the polysomes of the entire rough endoplasmic reticula (RER). (bar=200nm) (Cited from Ref.16).

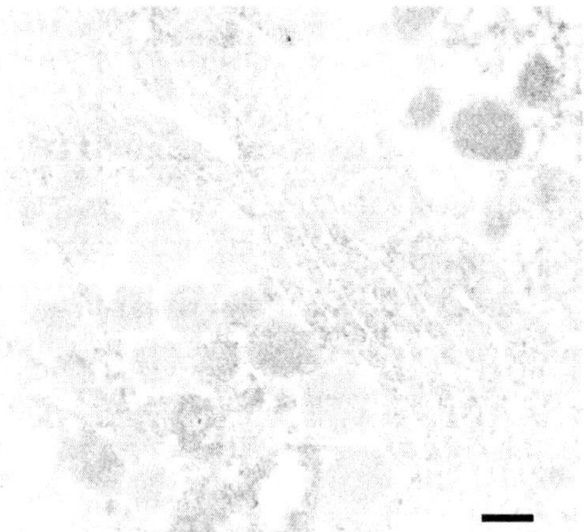

Figure. 2. Control study with a sense probe for GH mRNA. GH mRNA is not localized on the RER. (bar=200nm) (Cited from Ref.16).

EM-ISH: Postembedding Method

After fixation at 4°C overnight in 4% paraformaldehyde dissolved in PBS, anterior pituitary tissues were embedded in LR White resin (Polyscience; Warrington, PA, USA). They were carefully placed at the bottom of 00 gelatin capsules (Lilly Pharmaceuticals; Indianapolis, IN, USA), which were filled with LR White resin and sealed. After polymerization at 50°C for 24 hr in a vacuum oven, ultrathin sections were retrieved on Nickel grids. Hybridization was carried out on the grids at 37°C overnight after prehybridization at the same temperature for 30 min. The hybridization solution is the same as used for preembedding method. The concentration of the probe for rat GH mRNA was 1 ng/ml. After hybridization, the grids were dipped in 2xSSC, 1xSSC and then 0.5xSSC for 5 min, respectively. The hybridization signals were developed for 30 min with 20 nm streptavidin gold (British Biocell International, Cardiff, UK) diluted 1:50 in 1% BSA-PBS. After dipped in PBS and distilled water (DW) and dried at room temperature, the grids were inspected under electron microscopy. The control experiments carried out were hybridization with probes of sense or scramble sequence and without probes.

Hybridization signals for rat GH mRNA were localized on the polysomes of the RER using 20 nm streptavidin gold (figure 3). EM-ISH with a sense probe for rat GH mRNA showed no hybridization signals (figure 4). The hybridization signal intensity is lower than that for the preembedding method.

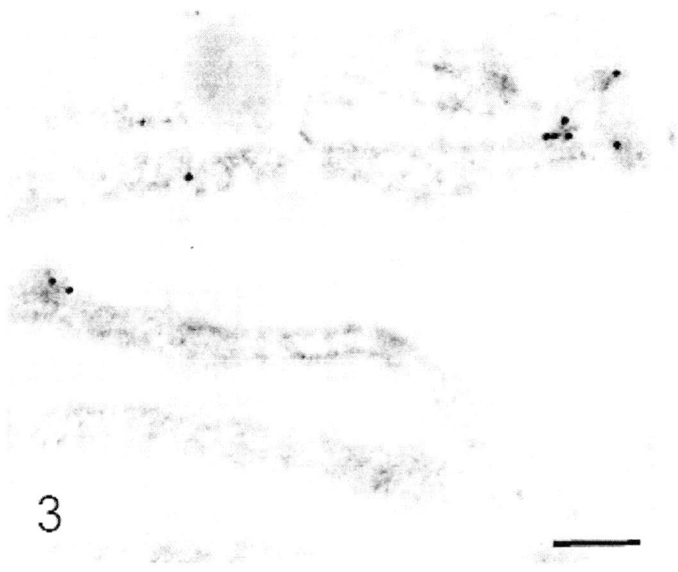

Figure 3. Hybridization signals for rat GH mRNA are localized on the polysomes of the RER using 20 nm streptavidin gold. (bar=200nm) The hybridization signal intensity is lower than that for the preembedding method. (Cited from Ref.16).

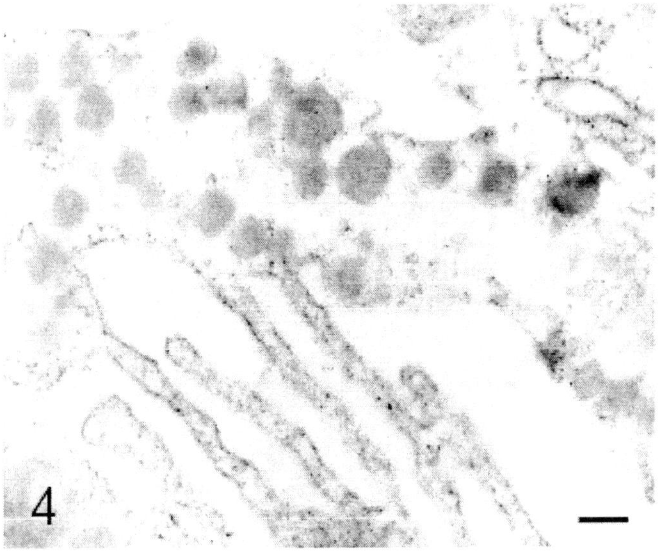

Figure 4. EM-ISH with a sense probe for rat GH mRNA shows no hybridization signals. (bar=200nm) (Cited from Ref.16).

Pathophysiological Studies on Changes in Ultrastructural Expression of PRL Mrna Induced by Estrogen and Bromocriptine: Preembedding and Postembedding Method

Female Wistar-Imamichi rats treated intramuscularly with 5 mg estradiol dipropionate (E2 depot: Ovahormon Depot; Teikoku Zoki Co. Ltd., Tokyo, Japan) every 4 weeks. They were sacrificed 7 weeks after injection with or without subcutaneous injection of 1 mg bromocriptine (Sandoz Pharmaceutical Co. Ltd., Basel, Switzerland) for 4 days. The anterior lobes removed from their pituitary glands were immediately fixed at 4 °C overnight in 4% paraformaldehyde dissolved in PBS and served for ultrastructural ISH studies.

EM-ISH studies revealed the whirling changes of the RER in the specimens of female rats given estrogen for 7 weeks and also frequent but focally localized hybridization signals of PRL mRNA on the polysomes of the whirling RER (figure.5). After bromocriptine administration, PRL mRNA expression at the light microscopic level decreased markedly, and electron microscopic examination revealed diffuse localization of PRL mRNA hybridization signals on the distorted, vesiculated, and partly dilated RER (figure.6). There were also increased number of secretory granules, which resulted in the increased PRL immunoreactivity induced by bromocriptine. Using the postembedding method, intense PRL mRNA hybridization signals were localized on the RER of specimens from female rats treated with estrogen for 7 weeks. After bromocriptine administration, the PRL mRNA hybridization signals dispersed and decreased.

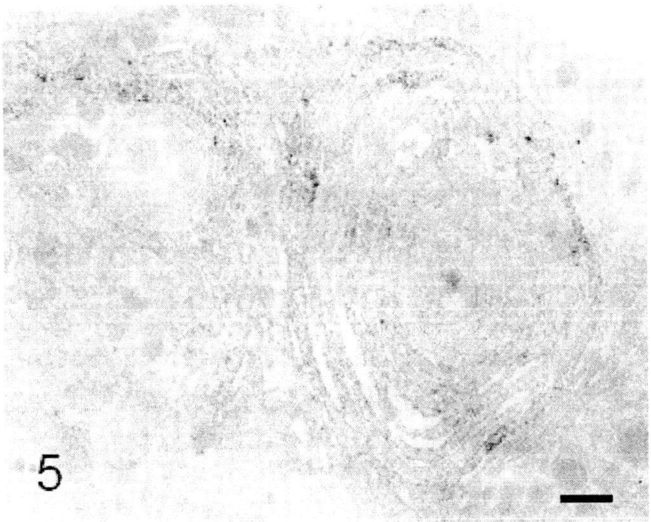

Figure 5. Photograph of pituitary cells from female rats given estrogen for 7 weeks subjected to preembedding electron microscopic ISH. PRL mRNA is localized frequently on the polysomes of the whirling RER. (bar=200nm).

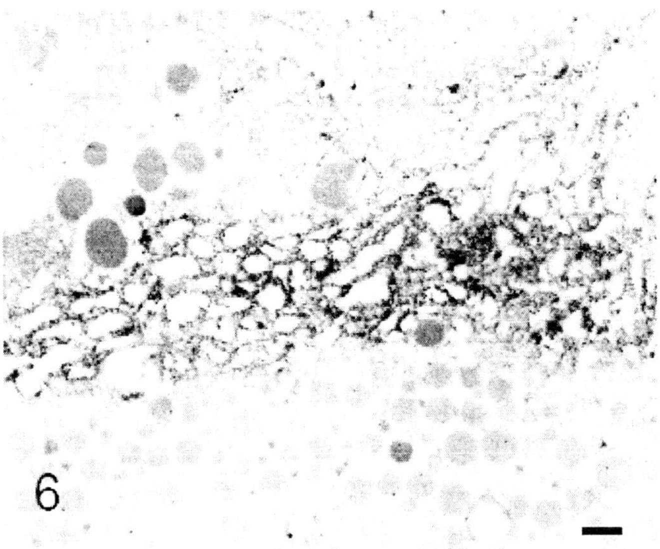

Figure 6. Electron microscopic observation of pituitary cells after bromocriptine administration reveals PRL mRNA hybridization signals localized on the distorted, vesiculated, and partly dilated RER, and also increased number of accumulated secretory granules. (bar=200nm) (Cited from Ref.16).

COMBINED ISH AND IHC AT AN ELECTRON MICROSCOPIC LEVEL

The hybridization procedure on 6 μm thick tissue specimens mounted on 3-aminopropylmethoxysilane-coated slides is the same as described as above. The hybridization signals of mRNA were detected with ABC-HRP, and then developed with DAB and H2O2. After osmification and dehydration with ethanol, tissue sections were embedded in Epon resin, which was polymerized at 60°C for 2 days. After polymerization, ultrathin sections were attached to Nickel grids. Subsequently, in order to retrieve the immunoreactivity of the targeted protein, the ultrathin sections embedded in Epon resin were etched with 10% H2O2 for 30 min or with 4% sodium periodate for 10 min, followed by washing with DW. Immunohistochemical staining for rat GH was carried out at room temperature for 1 hr using the antibody against rat GH (monkey, diluted 1:100 with PBS), supplied by the National Institute of Diabetes and Digestive and Kidney Diseases (NIDDK) (Bethesda, MD, USA). The grids were washed with 0.1 M phosphate buffer (PB), and the immunoreaction was visualized at room temperature for 1 hr with 20 nm protein A colloidal gold (E.Y. Laboratories Inc., San Mateo, CA, USA) diluted 1:40 in 0.1 M PB. After washed with 0.1 M PB and DW, the grids were dried at room temperature, and observed under electron microscopy. The immunohistochemical control experiments included substitution of normal monkey serum for anti-GH antibody.

As shown in the above section, EM-ISH with an antisense probe for rat GH mRNA revealed its diffuse localization on the polysomes of the entire RER. Subsequent immunohistochemical staining using anti-rat GH antibody and 20 nm protein A colloidal gold identified rat GH on the secretory granules (figure 7). Colloidal gold signals for GH were distributed mainly on the secretory granules, and were also identified in the cisternae of the RER (figure 7). The immunoreactivity retrieved after etching process with 10% H2O2 was similar and comparable to that with 4% sodium periodate. Immunohistochemical control experiments substituting normal monkey serum for anti-GH antibody combined with preembedding EM-ISH method revealed no reactions of protein A colloidal gold particles on the secretory granules.

The only flaw of this combined method is the deosmification and degradation of the signals of mRNA caused by etching process using H2O2 or sodium periodate. In order to solve this problem, we have recently used LR White resin for tissue embedment [15]. EM-ISH using LR White resin for tissue embedment also localized rat GH mRNA on the polysomes of the entire RER (figure 8). Subsequent immunohistochemical staining identified rat GH protein both on the secretory granules and in the cisternae of the RER (figure.8).

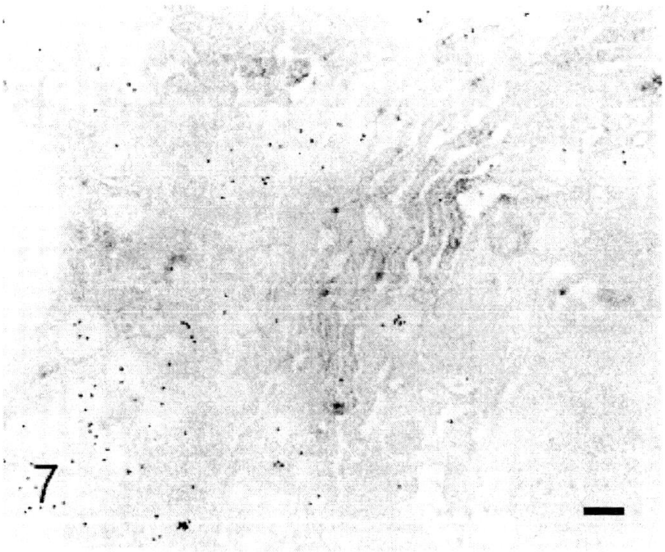

Figure 7. Combined immunohistochemistry and preembedding EM-ISH at the ultrastructural level. Positive signals for GH mRNA are localized as osmium black on the polysomes of the RER using biotinylated antisense oligonucleotide probe. GH is identified mainly on the secretory granules with 20 nm protein A colloidal gold particles (arrow), and is also identified in the cisternae of the RER (arrow head). (bar=200nm) (Cited from Ref.16).

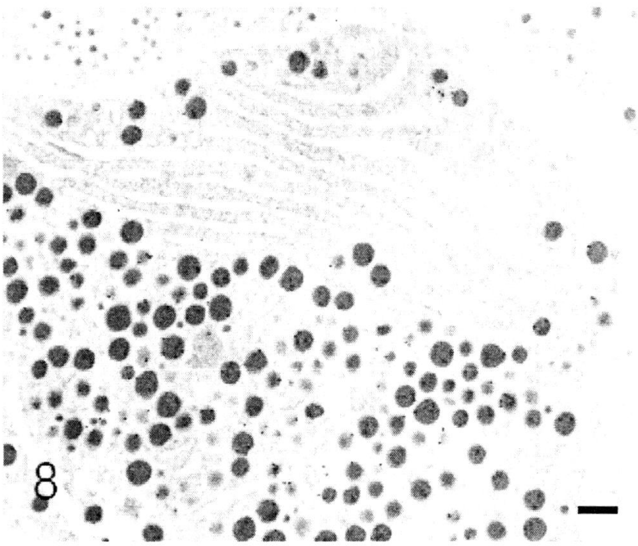

Figure 8. Using LR White resin for tissue embedment, rat GH mRNA is also localized on the polysomes of the entire rough endoplasmic reticulum. Subsequent immunohistochemical staining identifies rat GH protein both on the secretory granules (arrow) and in the cisternae of the rough endoplasmic reticulum (arrow head). (bar=200nm).

COMBINED ISH AND IHC USING QDOTS FOR THE DETECTION OF MRNA AND PROTEIN

Six μm thick tissue specimens were mounted on 3-aminopropylmethoxysilane-coated slides. After air drying for 1 h, tissue sections were washed with PBS for 15 min. Subsequently they were treated with 0.1 μg/ml proteinase K at 37°C for 30 min, followed by treatment with 0.25% acetic anhydride in 0.1 M triethanolamine for 10 min. The slides were washed in 2x SSC at room temperature for 3 min and then prehybridized at 37°C for 30 min. The biotinylated probe for rat GH or PRL mRNA with the concentration of 0.1 ng/μl was diluted with the prehybridization solution and hybridization was carried out overnight at 37°C. After hybridization, the slides were washed with 2xSSC, 1xSSC, and then 0.5xSSC for 15 min each at room temperature. Then the slides were washed in PBS, and immunohistochemical staining for rat GH or PRL was carried out for 1 h at room temperature. The antibodies used were anti-rat GH antibody (rabbit, polyclonal, 1:400 diluted in BSA - PBS, from Biogenesis, Ltd., Poole, Dorset, United Kingdom), or anti-rat PRL antibody (rabbit, polyclonal, diluted 1:100 with PBS, supplied by the National Institute of Diabetes and Digestive and Kidney Diseases (NIDDK), Bethesda, MD, U.S.A.). The slides were washed in PBS, and Qdot 655 conjugated with anti-rabbit IgG and Qdot 605 conjugated with streptavidin (Quantum Dot Corp., Hayward, CA, U.S.A.) were applied for 30 min at room temperature. Qdot 655 conjugated with anti-rabbit IgG was used for the detection of GH or PRL protein, and Qdot 605 conjugated with streptavidin was used for the detection of GH or PRL mRNA. The slides were washed in PBS, and nuclear staining was carried out with methyl green. They were inspected under CLSM.

The negative control experiments for ISH included hybridization studies with probes of sense or scramble sequence, as well as studies without probes. The negative control experiments for IHC involved the substitution of normal rabbit immunoglobulin fraction for primary antibodies.

With the confocal mode, GH protein was observed as a 655 nm emission signal using using Qdot 655 conjugated with anti-rabbit IgG (figure 9), and GH mRNA was observed as a 605 nm emission signal using Qdot 605 conjugated with streptavidin (figure 9). When GH mRNA and protein were located either in the same or adjacent places, their signals were detected in the mixed color image (figure 9). The mixed color images of GH mRNA and protein conjugated with different sized Qdots mean that these molecules are located within the three-dimensional distance of 25±13nm [45,46].

With the confocal mode, PRL protein was observed as a 655 nm emission signal using Qdot 655 (figure 10), and PRL mRNA was observed as a 605 nm emission signal using Qdot 605 (figure 10). Merged images of both signals revealed that when PRL mRNA and protein were located either in the same or adjacent places, their signals were detected in the mixed color image (figure10).

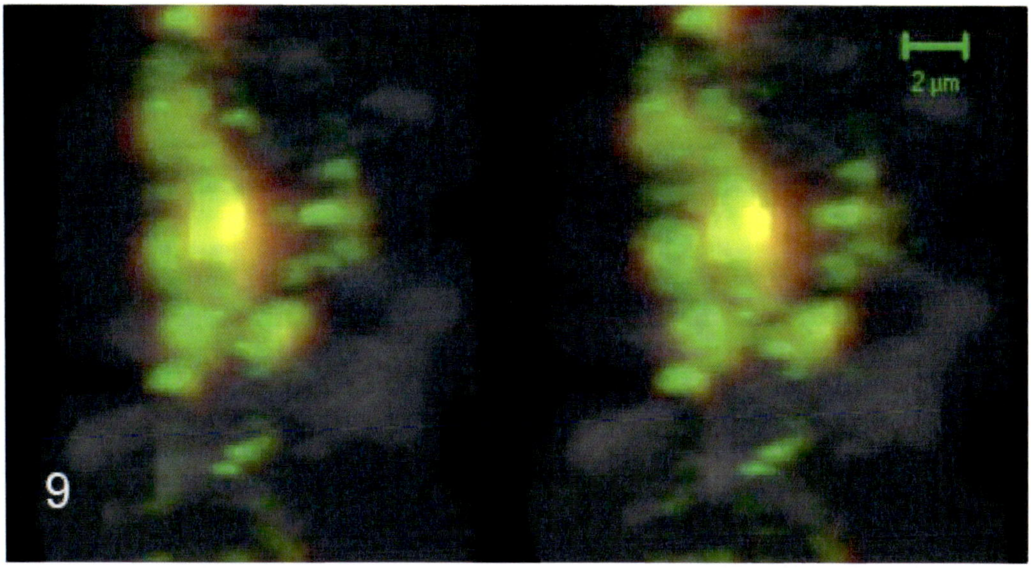

Figure 9. With the confocal mode, GH protein was observed as a 655 nm emission signal using Qdot 655 (red), and GH mRNA was observed as a 605 nm emission signal using Qdot 605 (green). When GH mRNA and protein were located in the same or adjacent places, their signals were detected in the mixed color images (yellow). (stereo-images, bar=2µm).

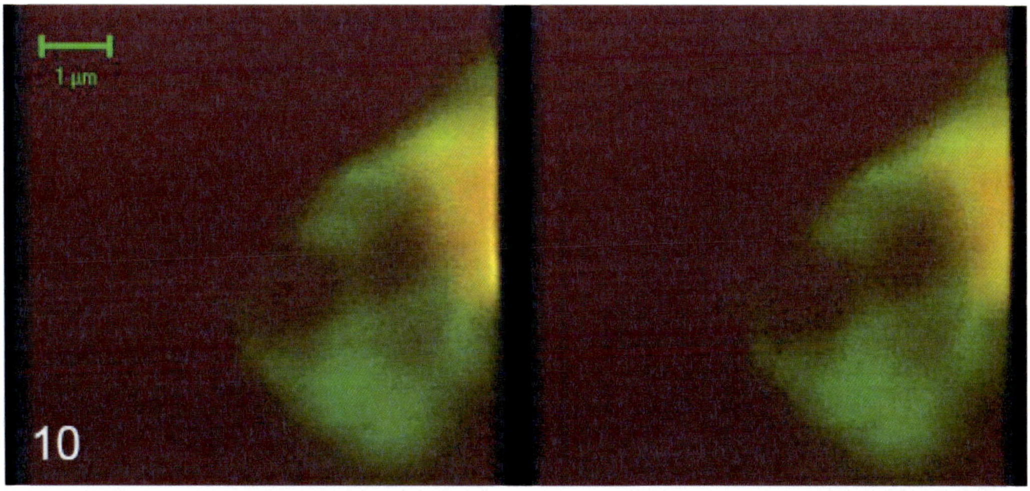

Figure 10. With the confocal mode, PRL protein was observed as a 655 nm emission signal using Qdot 655 (red), and PRL mRNA was observed as a 605 nm emission signal using Qdot 605 (green). When PRL mRNA and protein were located in the same or adjacent places, their signals were detected in the mixed color images (yellow). (stereo-images, bar=1□m).

Discussion

EM-ISH is essential for the visualization of the intracellular distribution and role of mRNA for protein synthesis. The major concerns of EM-ISH are to maintain tissue morphology and retain the messages. In order to obtain satisfactory morphological preservation, we have routinely utilized 6 μm thick frozen sections fixed in 4% paraformaldehyde for the preembedding method and tissues embedded in LR White resin for the postembedding method. In the preembedding method, both ultrastructure and mRNA are sufficiently preserved. The preembedding method has a benefit in that hybridization signals can be confirmed light microscopically. In the postembedding method using tissues embedded in LR White resin, ultrastructure is also sufficiently preserved. As described in our previous reports [6-8], compared with the preembedding method, the postembedding method has several difficulties: 1) Difficulty of message preservation during polymerization of LR White resin at high temperature for an extended period of time, which leads to mRNA degradation. 2) Possible non-specific signals due to the non-specific affinity of gold particles.

In this study we found the difference in localization of GH and PRL synthesis on the polysomes of the RER. Hybridization signals of GH mRNA are distributed diffusely on the entire RER, whereas those of PRL mRNA are scattered and distributed focally on the RER even after stimulated by estrogen. After bromocriptine treatment, hybridization signals of PRL mRNA are localized diffusely on the RER. Based on these findings, we speculate that some unknown factors can localize messenger RNA on RER. These may be related with signal recognition particles, their receptor, and microtubule transportation pathways [47]. This unknown factor will be influenced by several external or internal factors and will localize the site of protein synthesis on RER. The site of protein synthesis on RER may differ according to proteins. EM-ISH can serve the morphological and pathophysiological investigation on mRNA expression induced by some stimulatory factors. Further innovation will be required for these EM-ISH methods to provide adequate resolution for elucidating the intracellular localization of mRNA on free ribosomes.

Another of our concerns is to visualize mRNA and encoded protein simultaneously in the same cell. There seem to be two major problems to be resolved in this ultrastructural double staining method. One is to retain the messages and the other is to maintain the immunoreactivity of the encoded protein in the same cell. As stated above, as for mRNA preservation the preembedding EM-ISH method using frozen sections fixed in 4% paraformaldehyde has more advantages over the postembedding EM-ISH method using tissues embedded in LR White resin [6-8]. Frozen sections fixed in 4% paraformaldehyde has better morphological preservation than immediately frozen sections. Based on this assessment apropos of maintaining tissue morphology and retaining the messages, we utilize the preembedding EM-ISH method using 6 μm thick frozen sections fixed in 4% paraformaldehyde for the simultaneous detection of mRNA and encoded protein. The immunoreactivity can be retrieved by the etching process with H2O2 or sodium periodate even after modification such as osmification and embedment in Epon resin, and that tissues embedded in Epon resin can serve the ultrastructural simultaneous detection of messages and encoded proteins. The only problem is the deosmification and degradation of the signals of mRNA, which are caused by etching process with H2O2 or sodium periodate. In order to solve this problem, we have recently used LR White resin for tissue embedment [15]. In LR White resin-embedded tissues, retrieval of immunoreactivity with H2O2 or sodium periodate is not required, and therefore the gradation of the signals of mRNA can be avoided. To our knowledge, only five reports, except for our previous ones [14-16], describing the ultrastructural simultaneous detection of mRNA and encoded protein have been published, in each of which postembedding EM-ISH method using colloidal gold particle was utilized [48-52]. However, in these postembedding EM-ISH studies, the relatively frequent non-specific reactions of colloidal gold particles used for the detection of mRNA were observed in the cisternae of the RER. EM-ISH method for Lowicryl K4M-embedded tissues is generally supposed to have some difficulty in morphological preservation. The decreased message preservation was observed in the postembedding EM-ISH studies using LR White-embedded tissues [6,51]. From the viewpoint of specific hybridization signal detection and morphological preservation, the present combined immunohistochemistry and preembedding EM-ISH method is supposed to be superior to the previously reported postembedding EM-ISH method, and is successful and satisfactory ultrastructural simultaneous identification of mRNA and encoded protein. Translation of mRNA for secreted proteins such as GH is initiated on free ribosomes and then translocated to the polysomes on the RER with the aid of SRP once the signal peptide is produced. Synthesized proteins are secreted into the luminal space of the RER and subsequently transported to secretory granules via Golgi's apparatus. This ultrastructural double staining method for mRNA and encoded protein can provide an important clue for elucidating the intracellular correlation of mRNA translation and secretion of translated protein.

Qdots have narrow, symmetric emission spectra with multiple resolvable colors that can be excited simultaneously using a single excitation wavelength. The color of Qdots can be tuned to any chosen wavelength by simply changing their size. This property enables multiple labeling of subcellular molecules. Qdots have an advantage over conventional fluorophores, such as FITC and Texas Red, due to photostability and prominent signal intensity. Qdot signals were more than 11-fold stronger than those of fluorescein [41]. Using these properties, we successfully visualized the intracellular localization of pituitary hormone and mRNA by using different sized Qdots with CLSM [42,43]. This analysis has several merits: it can be used with light microscopic specimens; it can be observed in any chosen cells and any chosen depth of the section; it can reconstruct three-dimensional images. Three-dimensional images of the intracellular localization of mRNA and encoded protein enhance our three-dimensional understanding concerning the localization of mRNA and secreted protein. Two different sized Qdots can discriminate between two molecules that are located in the three-dimensional distance more than 25±13nm [45,46]. Therefore, the mixed color images of GH mRNA and protein conjugated with different sized Qdots mean that these molecules are located within the three-dimensional distance of 12-38nm.

Meanwhile, combined EM-ISHandIHC method provides higher resolutional images of subcellular organelles than combined ISH and IHC using Qdots and CLSM, and therefore, EM-ISHandIHC method is essential for analyses with high resolutions, although its images of mRNA and protein are two-dimensional.

By using ISH and IHC with Qdots and CLSM, one can optimize the visualization of the relationship between protein and mRNA simultaneously and three-dimensionally. It enables us to visualize functional images of proteins in relation with mRNA synthesis and localization. By using EM-ISHandIHC method, one can optimize the two-dimensional visualization of the relationship between protein and mRNA simultaneously in high resolutions. Then, both EM-ISHandIHC and ISH and IHC using Qdots and CLSM are useful for understanding the relationship between protein and mRNA simultaneously in two or three dimensions.

Conclusion

For two- or three- dimensional imagings of subcellular localization of pituitary hormone and its mRNA, both EM-ISHandIHC and CLSM observation using Qdots are essential.

References

[1] Guitten,y AF, and Bloch, B. (1989) Ultrastructural detection of the vasopressin messenger RNA in the normal and Brattleboro rat. *Histochemistry* 92:277-281.

[2] Wolber, RA, Beals, TF, and Maassab HF. (1989) Ultrastructural localization of Herpes simplex virus RNA by in situ hybridization. *J Histochem Cytochem* 37:97-104.

[3] Trembleau, A, Calas, A, and Fevre-Montange, M. (1990) Ultrastructural localization of oxytocin mRNA in the rat hypothalamus by in situ hybridization using a synthetic oligonucleotide. *Brain Res Mol Brain Res* 8:37-45.

[4] Pomeroy, ME, Lawrence, JB, Singer, RH, and Billings-Gagliardi, S. (1991) Distribution of myosin heavy chain mRNA in embryonic muscle tissue visualized by ultrastructural in situ hybridization. *Dev Biol* 143:58-67.

[5] Le Guellec, D, Trembleau, A, Pechoux, C, Gossard, F, and Morel, G. (1992) Ultrastructural non-radioactive in situ hybridization of GH mRNA in rat pituitary gland: preembedding vs ultrathin frozen sections vs postembedding. *J Histochem Cytochem* 40:979-986.

[6] Matsuno, A, Ohsugi, Y, Utsunomiya, H, Takekoshi, S, Osamura, RY, Watanabe, K, and Teramoto, A. (1994) Ultrastructural distribution of growth hormone, prolactin mRNA in normal rat pituitary cells: A comparison between preembedding and postembedding methods. *Histochemistry* 102:265-270.

[7] Matsuno, A, Teramoto, A, Takekoshi, S, Utsunomiya, H, Ohsugi, Y, Kishikawa, S, Osamura, RY, Kirino, T, and Lloyd RV. (1994) Application of biotinylated oligonucleotide probes to the detection of pituitary hormone mRNA using Northern blot analysis, in situ hybridization at light and electron microscopic levels. *Histochem J* 26:771-777.

[8] Matsuno, A, Ohsugi, Y, Utsunomiya, H, Takekoshi, S, Sanno, N, Osamura, RY, Watanabe, K, Teramoto, A, and Kirino, T. (1995) Changes in the ultrastructural distribution of prolactin and growth hormone mRNAs in pituitary cells of female rats after estrogen and bromocriptine treatment, studied using in situ hybridization with biotinylated oligonucleotide probes. *Histochem Cell Biol* 104:37-45.

[9] Morel, G, Chabot, JG, Gossard, F, and Heisler, S. (1989) Is atrial natriuretic peptide synthesized and internalized by gonadotrophs? *Endocrinology* 124:1703-1710.

[10] Morel, G, Dihl, F, and Gossard, F. (1989) Ultrastructural distribution of growth hormone (GH) mRNA and GH intron 1 sequences in rat pituitary gland: effects of GH releasing factor and somatostatin. *Mol Cell Endocrinol* 65:81-90.

[11] Jirikowski, GF, Sanna, PP, and Bloom, FE. (1990) mRNA coding for oxytocin is present in axons of the hypothalamo-neurohypophyseal tract. *Proc Natl Acad Sci USA* 87:7400-7404.

[12] Le Guellec, D, Frappart, L, and Willems, R. (1990) Ultrastructural localization of fibronectin mRNA in chick embryo by in situ hybridization using 35S or biotin labeled cDNA probes. *Biol Cell* 70:159-165.

[13] Le Guellec, D, Frappart, L, and Desprez, PY. (1991) Ultrastructural localization of mRNA encoding for the EGF receptor in human breast cell cancer line BT20 by in situ hybridization. *J Histochem Cytochem* 39:1-6.

[14] Matsuno, A, Utsunomiya, H, Ohsugi, Y, Takekoshi, S, Sanno, N, Osamura, RY, Nagao, K, Tamura, A, and Nagashima, T. (1996) Simultaneous ultrastructural identification of growth hormone and its messenger ribonucleic acid using combined immunohistochemistry and non-radioisotopic in situ hybridization: a technical note. *Histochem J* 28:703-707.

[15] Matsuno, A, Ohsugi, Y, Utsunomiya, H, Takekoshi, S, Munakata, S, Nagao, K, Osamura, RY, Tamura, A, and Nagashima, T. (1998) An improved ultrastructural double-staining method of rat growth hormone and its mRNA using LR White resin: a technical note. *Histochem J* 30:105-109.

[16] Matsuno, A, Nagashima, T, Osamura, RY, and Watanabe, K. (1998) Application of ultrastructural in situ hybridization combined with immunohistochemistry to pathophysiological studies of pituitary cell: Technical Review. Acta Histochem *Cytochem* 31:259-265.

[17] Matsuno, A, Nagashima, T, Takekoshi, S, Utsunomiya, H, Sanno, N, Osamura, RY, Watanabe, K, Tamura, A, and Teramoto, A. (1998) Ultrastructural simultaneous identification of growth hormone and its messenger ribonucleic acid. *Endocrine J* 1998, 45[Suppl]:S101-S104.

[18] Matsuno, A, Itoh, J, Osamura, RY, Watanabe, K, and Nagashima, T. (1999) Electron microscopic and confocal laser scanning microscopic observation of subcellular organelles and pituitary hormone mRNA: application of ultrastructural in situ hybridization and immunohistochemistry to the pathophysiological studies of pituitary cells. *Endocr Pathol* 10:199-211.

[19] Matsuno, A, Nagashima, T. Ohsugi, Y. Utsunomiya, H, Takekoshi, S, Munakata, S, Nagao, K, Osamura, RY, and Watanabe, K. (2000) Electron microscopic observation of intracellular expression of mRNA and its protein product: Technical review on ultrastructural in situ hybridization and its combination with immunohistochemistry. *Histol Histopathol* 15:261-268.

[20] Osamura. RY, Itoh. Y, and Matsuno, A. (2000), Application of plastic embedding to electron microscopic immunocytochemistry and in situ hybridization in observations of production and secretion of peptide hormones. *J Histochem Cytochem* 48:885-892.

[21] Osamura, RY, Tahara, S, Kurotani,, R, Sanno, N, Matsuno, A, and Teramoto, A. (2000) Contributions of immunohistochemistry and in situ hybridization to the functional analysis of pituitary adenomas. *J Histochem Cytochem* 48:445-458.

[22] Arndt-Jovin, DJ, Robert-Nicoud, M, Kaufman, SJ and Jovin, TM. (1985) Fluorescence digital imaging microscopy in cell biology. *Science* 230;247-256.

[23] Arndt-Jovin, DJ, Robert-Nicoud, M and Jovin, TM. (1990) Probing DNA structure and function with a multi-wavelength fluorescence confocal laser microscope. *J Microsc* 157;61-72.

[24] Bauman, JG, Bayer, JA, and van Dekken H. (1990) Fluorescent in-situ hybridization to detect cellular RNA by flow cytometry and confocal microscopy. *J Microsc* 157;73-81.

[25] Hozak, P, Novak, JT, and Smetana, K. (1989) Three-dimensional reconstructions of nucleolus-organizing regions in PHA-stimulated human lymphocytes. *Biol Cell* 66;225-233.

[26] Itoh, J, Osamura, RY, and Watanabe, K. (1992) Subcellular visualization of light microscopic specimens by laser scanning microscopy and computer analysis: a new application of image analysis. *J Histochem Cytochem* 40;955-967.

[27] Itoh, J, Sanno, N, Matsuno, A, Itoh, Y, Watanabe, K, and Osamura, RY. (1997) Application of confocal laser scanning microscopy (CLSM) to visualize prolactin (PRL) and PRL mRNA in the normal and estrogen-treated rat pituitary glands using non-fluorescent probes. *Microsc Res Tech* 39:157-167.

[28] Itoh, J, Matsuno, A, Yamamoto, Y, Kawai, K, Serizawa, A, Watanabe, K, Itoh, Y, and Osamura, RY. (2001) Confocal laser scanning microscopic imaging of subcellular organelles, mRNA, protein products, and the microvessel environment. *Acta Histochem Cytochem* 34:285-297.

[29] Michel, E, and Parsons, JA. (1990) Histochemical and immunocytochemical localization of prolactin receptors on Nb2 lymphoma cells: applications of confocal microscopy. *J Histochem Cytochem* 38;965-973.

[30] Robinson, JM, and Batten, BE. (1989) Detection of diaminobenzidine reactions using scanning laser confocal reflectance microscopy. *J Histochem Cytochem* 37;1761-1765.

[31] Takamatsu, T and Fujita, S. (1988) Microscopic tomography by laser scanning microscopy and its three-dimensional reconstruction. *J Microsc* 149;167-174.

[32] Tao, W, Walter, RJ, and Berns, MW. (1988) Laser-transected microtubules exhibit individuality of regrowth, however most free new ends of the microtubules are stable. *J Cell Biol* 107;1025-1035.

[33] White, JG, Amos, WB, and Fordham, M. (1987) An evaluation of confocal versus conventional imaging of biological structures by fluorescence light microscopy. *J Cell Biol* 105;41-48.

[34] Bruchez, M Jr., Moronne, M, Gin, P, Weiss, S, and Alivisatos, AP. (1988) Semiconductor nanocrystals as fluorescent biological labels. *Science* 281;2013-2016.

[35] Wang, C, Shim, M, and Guyot-Sionnest, P. (2001) Electrochromic nanocrystal quantum dots. *Science* 291;2390-2392.

[36] Chan, WC, Maxwell, DJ, Gao, X, Bailey, RE, Han, M, and Nie, S. (2002) Luminescent quantum dots for multiplexed biological detection and imaging. *Curr Opin Biotechnol* 13;40-46.

[37] Gao, X. Chan, WC, and Nie, S. (2002) Quantum-dot nanocrystals for ultrasensitive biological labeling and multicolor optical encoding. *J Biomed Opt* 7;532-537.

[38] Gao, X, and Nie, S. (2003) Molecular profiling of single cells and tissue specimens with quantum dots. *Trends Biotechnol* 21;371-373.

[39] Han, M, Gao, X, Su, JZ, and Nie, S. (2001) Quantum-dot-tagged microbeads for multiplexed optical coding of biomolecules. *Nat Biotechnol* 19;631-635.

[40] Pathak, S, Choi, SK, Arnheim, N, and Thompson, ME. (2001) Hydroxylated quantum dots as luminescent probes for in situ hybridization. *J Am Chem Soc* 123;4103-4104.

[41] Xiao, Y, and Barker, PE. (2004) Semiconductor nanocrystal probes for human metaphase chromosomes. *Nucleic Acids Res* 32;e28.

[42] Matsuno, A, Itoh, J, Takekoshi, S, Nagashima, T, and Osamura, RY. (2005) Three-dimensional imaging of the intracellular localization of growth hormone and prolactin and their mRNA using nanocrystal (Quantum dot) and confocal laser scanning microscopy techniques. *J Histochem Cytochem* 53:833-838.

[43] Matsuno, A, Itoh, J, Takekoshi, S, Nagashima, T, and Osamura, RY. (2005) Two- or three- dimensional imagings of simultaneous visualization of rat pituitary hormone and its mRNA: comparison between electron microscopy and confocal laser scanning microscopy with semiconductor nanocrystals (Quantum dots). *Acta. Histochem Cytochem* 38:253-256.

[44] Lloyd RV, Jin L, and Chandler WF. (1991) In situ hybridization in the study of pituitary tissues. *Path Res Pract* 187:552-555.

[45] Lacoste, TD, Michalet, X, Pinaud, F, Chemla, DS, Alivisatos, AP, and Weiss, S. (2000) Ultrahigh-resolution multicolor colocalization of single fluorescent probes. *Proc Natl Acad Sci U. S. A.* 97:9461-9466.

[46] Michalet, X, Pinaud, F, Lacoste, TD, Dahan, M, Bruchez, MP, Alivisatos, AP, and Weiss, S. (2001) Properties of fluorescent semiconductor nanocrystals and their application to biological labeling. *Single Mol* 4:261-276.

[47] Matsuno A, Takekoshi S, Sanno N, Utsunomiya H, Ohsugi Y, Saito N, Kanemitsu H, Tamura A, Nagashima T, Osamura RY, and Watanabe K. (1997) Modulation of protein kinases and microtubule-associated proteins and changes in ultrastructure in female rat pituitary cells; Effects of estrogen and bromocriptine. *J Histochem Cytochem* 45:805-813.

[48] Egger D, Troxier M, and Bienz K. (1994) Light and electron microscopic in situ hybridization: non-radioactive labeling and detection, double hybridization, and combined hybridization-immunocytochemistry. *J Histochem Cytochem* 42:815-822.

[49] Escaig-Haye F, Grigogiev V, Sharova I, Rudneva V, Buckrinskaya A, and Fournier JG. (1992) Ultrastructural localization of HIV-1 RNA and core proteins. Simultaneous visualization using double immunogold labelling after in situ hybridization and immunocytochemistry. *J Submicrosc Cytol Pathol* 24:437-443.

[50] Gingras D, and Bendayan M. (1995) Colloidal gold electron microscopic in situ hybridization: combination with immunocytochemistry for the study of insulin and amylase secretion. *Cell Vision* 2:218-225.

[51] Morey AL, Ferguson DJP, and Fleming KA. (1995) Combined immunocytochemistry and non-isotopic in situ hybridization for the ultrastructural investigation of human parvovirus B19 infection. *Histochem J* 27:46-53.

[52] Singer RH, Langevin GL, and Lawrence, JB. (1989) Ultrastructural visualization of cytoskeletal mRNAs and their associated proteins using double-label in situ hybridization. *J Cell Biol* 108:2343-2353.

In: New Messenger RNA Research Communications
Editor: Lee B. Kwang, pp. 87-98

ISBN 978-1-60021-488-2

Chapter VI

A Novel Assay to Quantify Serum Messenger RNAs as a Tumor Marker and its Clinical Significance in Hepatocellular Carcinoma

Norimasa Miura[5 1], Junichi Hasegawa[1] and Goshi Shiota[2]

[1]Division of Pharmacotherapeutics, Department of Pathophysiological and Therapeutic Science, Faculty of Medicine, Tottori University, 86 Nishicho, Yonago, Tottori 683-8503, Japan.

[2]Division of Molecular and Genetic Medicine, Department of Genetic Medicine and Regenerative Therapeutics, Graduate School of Medicine, Tottori University, Yonago 683-8503, Japan

Abstract

Currently available tumor markers for hepatocellular carcinoma (HCC) are α-fetoprotein (AFP), lens culinaris agglutinin-reactive AFP (AFP-L3), and des-γ-carboxy prothrombin (DCP). However, their diagnostic potential cannot surpass abdominal ultrasonography (US) as modalities to detect small HCC at an early stage, resulting in a potential delay in its diagnosis. There is a need to develop an alternative sensitive marker to improve the early detection of HCC. We here introduce a newly developed quantitative detection method for serum hTERT mRNA, which has clinical significance in HCC diagnosis.

In 64 patients with HCC, 20 with liver cirrhosis, 20 with chronic hepatitis, and 50 healthy individuals, we measured serum hTERT mRNA by using the newly developed real-time quantitative RT-PCR with SYBR Green I. Briefly, we examined its sensitivity and specificity in HCC diagnosis, clinical significance in comparison with other tumor

[5] Corresponding author. Tel.: +81-859-38-6172; fax: +81-859-38-6170.E-mail address: mnmiura@grape.med.tottori-u.ac.jp (N. Miura)

markers, and its correlations with the clinical parameters by using multivariate analyses and the Friedman test.

Serum hTERT mRNA levels were higher in patients with HCC than those with chronic liver diseases. We demonstrated that hTERT mRNA expression was independently correlated with clinical parameters such as tumor size, number and degree of differentiation ($P<0.001$ in each). The sensitivity/specificity of hTERT mRNA in HCC diagnosis showed a 88.2%/70.0% ratio. hTERT mRNA proved, as expected, to be superior to AFP mRNA , AFP and DCP in HCC diagnosis. More importantly, hTERT mRNA in serum was correlated with that in HCC tissue.

We report in this review that serum hTERT mRNA is a novel and available marker for HCC detection. We are also carrying out a large-scale study regarding HCC diagnosis. Furthermore, this method will be applicable for a number of hTERT-positive malignancies, and for the quantification of the expression of RNA genes or genes with a weak protein expression level.

Key words: hTERT, real-time RT-PCR, cancer diagnosis, hepatocellular carcinoma, tumor marker

Abbreviations

hTERT, human telomerase reverse transcriptase protein;
HCC, hepatocellular carcinoma;
hTR, human telomerase RNA template;
HCV, hepatitis C virus;
HBV, hepatitis B virus;
AH, adenomatous hyperplasia;
AAH, atypical adenomatous hyperplasia;
LC, liver cirrhosis;
CH, chronic hepatitis;
AFP, α-fetoprotein; DCP;
des-γ-carboxy prothrombin;
ALT, alanine aminotransferase;
Alb, albumin.

Introduction

HCC is a highly fatal cancer that affects approximately half a million people worldwide [1, 2]. The incidence of HCC has been rising rapidly and this increase has generated great concerns among clinicians, researchers, and policy makers. The major risk factors of HCC that are relatively common include hepatitis C virus (HCV), hepatitis B virus (HBV), nonalcoholic steatohepatitis (NASH), and heavy alcohol consumption. Development of HCC is generally thought to depend on the rapid and long-term repetition of cell turnovers, indicating that HCC expands clonally from the hepatocytes which have suffered a long period of intense inflammation and in which senescence has been introduced. There exists a sense of urgency to detect an evidence of clonal occurrence of HCC at an early stage using a newly developed modality. That solution is provided here.

Tumor Markers for HCC

AFP has been used as a serum marker for HCC in humans for many years and has a 39%-65% sensitivity and a 76%-94% specificity [3]. However, we can improve the detection rate using ultrasonography (US) [4]. Several biomarkers such as DCP, AFP-L3 [5], human hepatocyte growth factor (HGF), and insulin-like growth factor-1(IGF-1), are promising, but none of these markers have been sufficiently validated for clinical use as well as AFP. Thus, there is an urgent need for novel biomarkers for the detection of early HCC [6]. The marker should be specific for HCC, enabling its detection at an early stage, easily measurable, and the test should be reproducible, minimally invasive, and acceptable to patients and physicians [7]. Among biomarkers identified at this stage, DCP and AFP-L3 have been most extensively studied and seem to be promising. DCP has a sensitivity and a specificity that ranges from 28% to 89% and from 87% to 96%, respectively, while for AFP-L3 this ranges from 36% to 96% and from 89% to 94%, respectively. Most of these markers are mainly measured by enzyme immunoassay (EIA) to detect a protein due to the benefit of its stability.

History of Circulating Nucleic Acids

Recently, a new, alternative candidate to protein has appeared on the diagnostic assay scene. A simple blood test for cancer detection has been the quest of many researchers. The finding that cell-free circulating nucleic acids in the plasma/serum of cancer patients has characteristics of tumor-derived nucleic acids [8] has invoked much interest. Nucleic acids can be found in small amounts in healthy and diseased human plasma/serum [9]. Although it had been well known that higher concentrations of DNA are present in the plasma of cancer patients sharing some characteristics with the DNA of tumor cells [10, 11], it was reported that mRNAs detected in the blood may reflect early events in a clinical condition. Since RNA in plasma/serum may be a suitable source for the development of non-invasive diagnostic, prognostic, and follow-up tests for cancer, this discovery provided us with very promising assays for the early detection of malignancies.

Circulating RNA for Diagnostic Source

A recent development in this new field is the finding of tumor-related RNA in the plasma/serum of cancer patients [12]. These features include tyrosine kinase mRNA [13], telomerase components [14, 15], the mRNAs that are encoded by different tumor-related genes [16, 17, 18, 19, 20], and viral mRNA [21]. As more RNA markers can be detected in the circulation of cancer patients than DNA markers, an assay using RNA markers turned out to have a higher sensitivity than other conventional assays. In one study, two telomerase markers in breast cancer yielded a 44% positive rate [14]. Nevertheless, telomerase RNA seems to be a promising marker since it can be found even in the serum of patients with small, undifferentiated breast cancers without any metastatic lesions. Dasi et al. showed that circulating telomerase RNA is a sensitive marker, using real-time reverse transcription-PCR (RT-PCR) [15]. In their study, 8 out of 9 plasma samples from colorectal cancer patients and 9 out of 9 plasmas from patients with lymphoma were positive for human telomerase reverse transcriptase; tests were negative in all 10 healthy individuals.

An Application of the Htert Mrnas in Blood for Diagnosis

The hTERT mRNA or endogeneous RNA component derived from cancer cells did not previously appear to be detectable in serum due to its instability by RNase present in the serum. However, since RNAs in the serum are stable at least for 24 hour after drawing blood [22, 23], it has been suggested that they can be detected, even in blood [24, 25]. Actually, hTERT mRNA can be detected in serum from breast cancer patients and its maximum sensitivity and specificity were at most 40% and 100%, respectively [26]. We previously reported that approximately 88% was qualitatively positive in the detection of HCC-derived RNA in serum [27], indicating that serum RNA can be applicable for diagnosis of other cancers as well as HCC. Thus, we singled out hTERT mRNA as the most appropriate molecule for cancer diagnosis. In this context, we focused on HCC, since hTERT mRNA gradually increases during the multistep process of hepatocarcinogenesis [28].

Materials, Methods, and the Results of HTERT mRNA in HCC

One hundred and four consecutive patients [64 patients with HCC, 20 with liver cirrhosis (LC), and 20 with chronic hepatitis (CH)] were enrolled in this study. All the HCC patients had LC as the underlying liver disease. The mean ages of patients with HCC, LC, and CH were 66, 64, and 55 yr, respectively. Sixty-six patients were infected with HCV, 30 with HBV, 3 with both viruses and 5 with no viral markers. The clinicopathological findings such as gender, age, etiology, Pugh score, Child classification, underlying liver disease, total bilirubin (TB), albumin (Alb), alanine aminotransferase (ALT), AFP, AFP-L3, DCP, HCV titer, HCV subtype, tumor number, tumor size, differentiation degree of tumor, and presence of metastasis were evaluated. Fifty healthy individuals including 12 females (from 22 - 83 yr old, mean age 58 yr) served as controls. To study whether hTERT mRNA in serum originates and is released from HCC in the liver, we harvested the surgically resected HCC tissues and sera in another 10 HCC patients. RNA was extracted with DNase treatment after three centrifugation steps of serum as we reported previously [26, 27]. The quantitative real-time RT-PCR was performed using SYBR Green I using a LightCycler with reproducibility. The dynamic ranges of real-time PCR analysis for hTERT mRNA and AFP mRNA were more than approximately 5 copies in this assay and we could exclude the possibility of false negativity in serum samples from patients with CH, LC, and controls. Both hTERT mRNA and AFP mRNA expression showed stepwise up-regulation with disease progression and the quantification was significantly higher in HCC than in LC, CH, and healthy individuals (P<0.0001, P<0.0001, and P<0.0001 in hTERT; P=0.011, P=0.044, and P<0.0001 in AFP, figure 1-A). To examine significant clinicopathological findings affecting hTERT and other markers, the significant difference among liver diseases and stratified categories in each clinical parameter was statistically evaluated. Alb, tumor size, the number of tumors, the differentiation degree of tumors and the presence of metastasis were significantly associated with hTERT mRNA expression (P<0.05, P<0.01, P<0.0001, P<0.0001, and P<0.05, respectively, table 1).

Table 1. Multivariate analysis of comparison of tumor markers with clinicopathological findings.

Clinical parameters	# of patient	Multivariate analyses and Freidman test			
		hTERT mRNA	AFP mRNA	AFP(-L3)	DCP
		p	*p*	*p*	*p*
Age mean:59 years old (range 22 to 83)		0.408	0.798	0.681 (0.690)	0.981
Gender M	94	0.761	0.089	0.412 (0.408)	0.380
F	60				
Etiology HBV	30	0.250	0.651	0.304 (0.052)	0.842
HCV	66				
HBV+HCV	3				
NBNC	2				
Alcohol	3				
Underlying lesion		0.060	0.973	0.621 (0.026)	0.534
Normal	50				
CH	32				
LC	70				
Albumin (g/dl)		0.018	0.340	0.540 (0.601)	0.001
Total bilirubin (mg/dl)		0.928	0.693	0.111 (0.432)	0.933
Alanine aminotransferase (IU/l)		0.538	0.149	0.001 (0.001)	0.978
Child-Pugh Scale A	21	0.136	0.573	0.373 (0.020)	0.001
B	44				
C	5				
AFP (ng/ml)		0.201	0.319	-	-
AFP-L3 (%)		0.123	0.425	-	-
DCP (mAU/ml)		0.854	0.651	-	-
Size of tumor (mm) < 20	18	<0.001	0.061	0.358 (0.001)	0.258
20~30	26				
> 30	20				
Number of tumors 1	10	<0.001	0.123	0.200 (0.012)	0.086
2	27				
? 3	27				
Differentiation degree		0.010	0.096	0.011 (0.010)	0.285
Well	33				
Moderate	27				
Undifferentiated	1				

Only hTERT mRNA correlated with albumin, tumor size, tumor number, and differentiation degree of tumor independently during the progression from chronic liver diseases to HCC. HBV: hepatitis B virus, HCV: hepatitis C virus, NBNC: non -HBV non-HCV, AH: adenomatous hyperplasia.

To examine correlations among biomarkers, we calculated Pearson's relative test and found that the hTERT mRNA level was significantly associated with the AFP mRNA level (P<0.05). To assess the accuracy of diagnostic tests, the matched data sets (chronic liver diseases patients and HCC patients) regarding biomarkers were analyzed by using receiver operator characteristic (ROC) curve analysis. This assay demonstrated a strong linear relation between copy number and PCR cycles using RNA controls (r^2>0.99). A correlation of hTERT mRNA between HCC tissue and serum was analyzed to examine whether serum hTERT mRNA is derived from original HCC tissues by using both the Paired t-test and Spearman's test. A significant correlation of hTERT mRNA in HCC tissues with that in serum was observed (P<0.01 and P<0.05, respectively, r=0.546, figure 1-B). DCP, AFP level, and AFP-L3 showed a significant correlation with tumor size (P<0.05, each) when it was stratified as <20 mm, 20-30 mm, and >30 mm in diameter. Furthermore, hTERT mRNA expression was closely associated with a well- and moderate- degree of differentiation of the HCC (P<0.05). hTERT mRNA was superior to any other tumor marker in differentiating HCC from chronic liver disease by the Friedrich test (P<0.01). ROC curve analyses showed that the sensitivity/specificity of hTERT mRNA for HCC was 88.2%/68.7% figure. 1-C, figure 2-A). Optimal predictive cut-off values for both mRNA expressions were 12,500 copies/0.2 ml and 3,000 copies/0.2 ml, respectively. The sensitivity/specificity in each tumor marker during hepatocarcinogenesis in the present assay is shown in figure 2-B. The positive predictive value (PPV)/negative predictive value (NPV) during hepatocarcinogenesis was 0.862/0.870 in hTERT mRNA. The PPV/NPV in AFP mRNA, AFP level, AFP-L3 and DCP was 0.695/0.741, 0.812/0.389, 0.778/0.277, and 0.852/0.405, respectively.

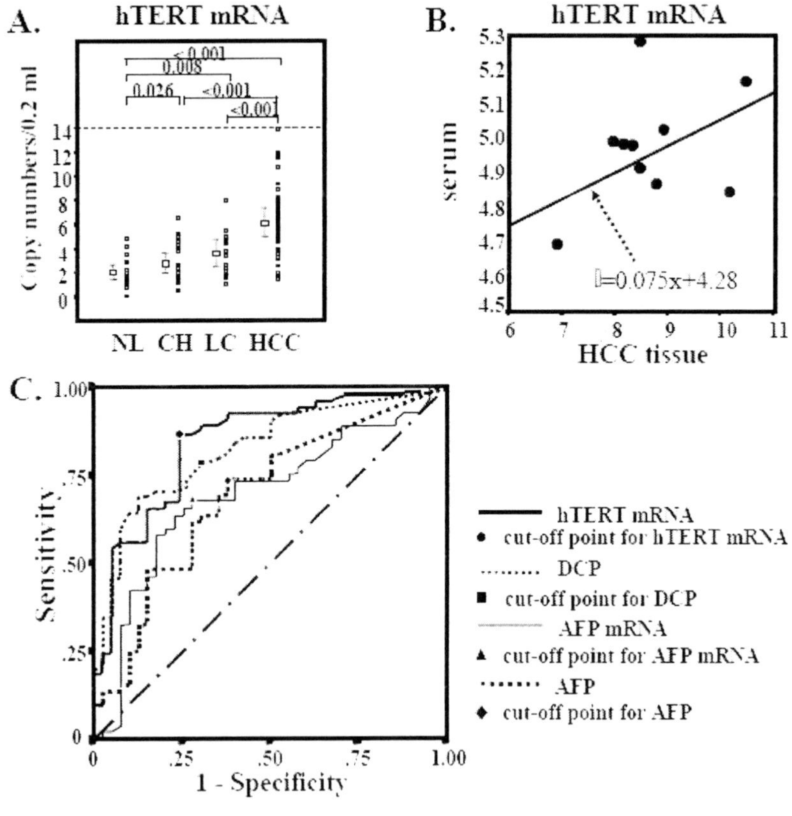

Figure 1.hTERT mRNA levels and AFP mRNA level (on logarithmic scales) in serum from patients with HCC, LC, CH, and healthy individuals by real-time RT-PCR. The 95% confidence interval in each group is shown beside the dots. Significant differences between 5 groups are shown in the upper part of the figure. NL: individual with normal liver, OL: other liver diseases, CH: chronic hepatitis, LC: liver cirrhosis, HCC: hepatocellular carcinoma (figure 1-A). Dot plot represents the significant correlation of hTERT mRNA level in serum with in that in HCC tissues in 10 patients. Those data were analyzed by the Paired t test (P=0.01) and the non-parametric Spearman's test (P=0.017) (figure 1-B).Receiver operator characteristic (ROC) curve analysis of hTERT mRNA and AFP mRNA expression. The curves shown were obtained by importing quantified raw data into SPSS13.0. The dotted line, bold dotted line, solid line, and bold solid line correspond to DCP, AFP, AFP mRNA, and hTERT mRNA, respectively. Each line has a cut-off point for a marker (figure 1-C).

A.

	Sensitivity	Specificity	*p* value	PPV/NPV
AFP	0.693	0.600	0.002	0.812/0.389
AFP-L3	0.563	0.925	0.304	0.778/0.277
DCP	0.815	0.635	< 0.001	0.852/0.405
AFP mRNA	0.716	0.675	< 0.001	0.695/0.741
hTERT mRNA	0.882	0.700	< 0.001	0.862/0.870

B.

Tumor marker	CH/LC (n=40) → HCC (n=64)	LC (n=20) → HCC (n=64)
AFP level	0.376	0.532
AFP-L3	0.144	0.228
DCP	0.317	0.479
AFP mRNA	0.001	0.001
hTERT mRNA	< 0.001	< 0.001

Figure 2.The sensitivity/specificity and PPV/NPV of each tumor marker for hepatocellular carcinoma are shown. PPV: positive predictive value, NPV: negative predictive value (figure 2-A). Statistical significance in each tumor marker during hepatocarcinogenesis is shown figure 2-B).

Considerations of Some Cases Below the Predictive Cut-Off Value of hTERT mRNA

In the present study, 8 out of 64 HCC patients were negative below the calculated predictive cut-off value for serum hTERT mRNA. Although the reason why hTERT mRNA was negative in these patients is not clear, 5 out of 8 hTERT mRNA-negative HCC patients had decompensated liver cirrhosis as the underlying disease. It was reported that decompensated liver cirrhosis had higher levels of serum TGF-β which promotes apoptosis of immortalized hepatocytes [29]. Therefore, in these cases, elevated TGF-β may stimulate apoptosis of immortalized cells, resulting in the reduction of hTERT mRNA [24]. In the other 3 hTERT mRNA-negative patients, HCC did not progress for a year. In 5 HCC patients, hTERT mRNA was followed for a year after resection of their HCC. The two patients whose HCC relapsed showed increased levels of serum hTERT mRNA. The hTERT mRNA levels in two out of the other 3 patients whose HCC did not relapse remained unchanged. The one patient whose HCC did not relapse showed an increase in the level of hTERT mRNA, and although the reason for this increase was not clear, underlying severe hepatitis may have affected this result.

Investigations into other Disease-Specific mRNAs and Their Clinical Significance

In other malignancies such as lung and ovarian cancer, this assay resulted in a 89.0% and 72.7%, and 90.9% and 90.0% sensitivity and specificity of hTERT mRNA, respectively (submitted for publication). We are searching for other disease-specific mRNAs as biomarkers to be applied in clinics or in primary care. The field of medical care we are engaged in is the diagnosis of intractable diseases (Table 2) and we think that it is necessary to find more informative mRNA such as hTERT rather than the presently widely used Positron Emission Tomography (PET), for use as a diagnostic vehicle. Furthermore, we are now studying inflammatory diseases (manuscript in preparation) or lifestyle-related diseases which are difficult to diagnose or whose clinical condition is difficult to estimate.

Conclusion

In a subsequent quantitative study, we not only improved the sensitivity of detection of unstable nucleotides in blood by removing cellular proteins and minimizing the contamination of cellular nucleic acids in serum, but also devised a primer set which can amplify hTERT mRNA efficiently [30]. Furthermore, the correlation of hTERT mRNA between tumor tissue and serum was demonstrated (Fig. 1-B), suggesting that hTERT mRNA detected in serum is derived from tumor cells. AFP is being widely used as a reliable marker of HCC, not at an earlier stage but rather at an advanced stage [31]. Since HCC repeatedly reccurs polyclonally after any treatment (its biological characteristic), the measurement of serum hTERT mRNA makes it possible to diagnose the reccurence, whose rate is remarkably high.

In this respect, we have to undergo a follow-up study after the treatment of HCC. hTERT mRNA expression was found to be closely associated with a well to moderate degree of differentiation of the HCC. Nakashio et al. previously reported the significant correlation of HCC differentiation with telomerase expression [32]. The results in the present study confirmed their findings. hTERT mRNA showed more sensitivity and specificity compared with AFP mRNA in HCC diagnosis. The higher specificity of hTERT mRNA in HCC may be related to the fact that AFP mRNA is produced in HCC cells and injured hepatocytes while hTERT is mainly produced in HCC cells. Waguri et al. proved that hTERT mRNA in circulating cancer cells derived from HCC tissues can be detectable [33]. The present study suggests that quantification of hTERT mRNAs in serum has diagnostic implications for HCC. We will evaluate the correlation of prognosis with hTERT mRNA [34] and assess the availability of hTERT mRNA in other cancers by comparison of hTERT mRNA with other tumor markers. We are performing a large-scale study, in which more than 500 patients are enrolled in this study and are being analyzed, confirming our results for monitoring and detecting HCC (unpublished data).

Table 2 A clinical application by this detection method using mRNAs circulating in blood.

1. Applications for malignancies	
a) Diagnosis of malignancies	
Lung cancer	**Adenocarcinoma**
	Squamous cell carcinoma
	Small cell lung carcinoma
	Large cell lung carcinoma
Gynecological malignancies	**Ovarian cancer**
	Uterine cancer
Gastroenterological malignancies	
	Hepatocellular carcinoma
	Stomach cancer
	Colon cancer
	Pancreatic cancer
	Esophageal cancer
Breast cancer	
Thyroid cancer	
Otolaryngological cancer	
Urinary cancer	
Sarcoma	
b) A Comparison of hTERT mRNA with PET in periodic medical examination for cancer	
c) An evaluation of the induction or effect of anticancer therapy	
2. Applications for other diseases	
a) Inflammatory diseases	**Fulminant hepatitis**
	Autoimmune disease
	Nonalcoholic steatohepatitis
b) Ischemic diseases	

Furthermore, this method will be applicable in the diagnosis of a number of hTERT-positive malignancies and of cancer-specific RNA genes or other genes showing a weak protein level. The systematic introduction of this assay at the primary care level can be expected in the near future.

References

[1] El-Serag HB, Mason AC. Rising incidence of hepatocellular carcinoma in the United States. *N Engl J Med* 1999; 340:745-50.

[2] Shirabe K, Takenaka K, Taketomi A, Kawahara N, Yamamoto K, Shimada M, et al. Postoperative hepatitis status as a significant risk factor for recurrence in cirrhotic patients with small hepatocellular carcinoma. *Cancer* 1996; 15(77):1050-5.

[3] Collier J, Sherman M. Screening of hepatocellular carcinoma. *Hepatology* 1998; 37:273-8.

[4] Dohmen K, Shirahama M, Onohama S, Miyamaoto Y, Torii Y, Irie K, et al. Differences in survival based on the type of follow-up for the detection of hepatocellular carcinoma: an analysis of 547 patients. *Hepatol Res* 2000; 18:110-21.

[5] Okuda H, Nakanishi T, Takatsu K, Saito A, Hayashi N, Yamamoto M, Takasaki K, Nakano M. Clinicopathologic features of patients with hepatocellular carcinoma seropositive for a-fetoprotein-L3 and seronegative for 12 des-γ-carboxyprothrombin in comparison with those seropositive for des-γ-carboxy prothrombin alone. *J Gastroenterol Hepatol* 2002; 17:772-8.

[6] Srivastava S, Gopal-Srivastava R. Biomarkers in cancer screening: a public health perspective. *J Nutr* 2002; 132(8 Suppl):2471S-5S. Review.

[7] Srinivas PR, Kramer BS, Srivastava S. Trends in biomarker research for cancer detection. *Lancet Oncol* 2001; 2:698-704. Review.

[8] Leon SA, Shapiro B, Sklaroff DM, Yaros MJ. Free DNA in the serum of cancer patients and the effect of therapy. *Cancer Res* 1977; 37:646-50.

[9] Stroun M, Anker P, Maurice P, Lyautey J, Lederrey C, Beljanski M. Neoplastic characteristics of the DNA found in the plasma of cancer patients. *Oncology* 1989; 46:318-22.

[10] Anker P, Mulcahy H, Chen XQ, Stroun M. Detection of circulating tumour DNA in the blood (plasma/serum) of cancer patients. *Cancer Metastasis Rev* 1999; 18:65-73. Review.

[11] Lo YM, Chan LY, Lo KW, Leung SF, Zhang J, Chan AT, Lee JC, Hjelm NM, Johnson PJ, Huang DP. Quantitative analysis of cell-free Epstein-Barr virus DNA in plasma of patients with nasopharyngeal carcinoma. *Cancer Res* 1999; 59:1188-91.

[12] Silva J, Silva JM, Garcia V, Garcia JM, Dominguez G, Bonilla F. RNA is more sensitive than DNA in identification of breast cancer patients bearing tumor nucleic acids in plasma. *Genes Chromosomes Cancer* 2002; 35:375-6.

[13] Kopreski MS, Benko FA, Kwak LW, Gocke CD. Detection of tumor messenger RNA in the serum of patients with malignant melanoma. *Clin Cancer Res* 1999; 5:1961-5.

[14] Chen XQ, Bonnefoi H, Pelte MF, Lyautey J, Lederrey C, Movarekhi S, Schaeffer P, Mulcahy HE, Meyer P, Stroun M, Anker P. Telomerase RNA as a detection marker in the serum of breast cancer patients. *Clin Cancer Res* 2000; 6:3823-6.

[15] Dasi F, Lledo S, Garcia-Granero E, Ripoll R, Marugan M, Tormo M, Garcia-Conde J, Alino SF. Real-time quantification in plasma of human telomerase reverse transcriptase (hTERT) mRNA: a simple blood test to monitor disease in cancer patients. *Lab Invest* 2001; 81:767-9.

[16] Silva JM, Dominguez G, Silva J, Garcia JM, Sanchez A, Rodriguez O, Provencio M, Espana P, Bonilla F. Detection of epithelial messenger RNA in the plasma of breast cancer patients is associated with poor prognosis tumor characteristics. *Clin Cancer Res* 2001; 7:2821-5.

[17] Kopreski MS, Benko FA, Gocke CD. Circulating RNA as a tumor marker: detection of 5T4 mRNA in breast and lung cancer patient serum. *Ann N Y Acad Sci* 2001; 945:172-8.

[18] Gal S, Fidler C, Lo YM, Chin K, Moore J, Harris AL, Wainscoat JS. Detection of gammaglobin mRNA in the plasma of breast cancer patients. *Ann N Y Acad Sci* 2001; 945:192-4.

[19] Fleischhacker M, Beinert T, Ermitsch M, Seferi D, Possinger K, Engelmann C, Jandrig B. Detection of amplifiable messenger RNA in the serum of patients with lung cancer. *Ann N Y Acad Sci* 2001; 945:179-88.

[20] Hasselmann DO, Rappl G, Rossler M, Ugurel S, Tilgen W, Reinhold U. Detection of tumor-associated circulating mRNA in serum, plasma and blood cells from patients with disseminated malignant melanoma. *Oncol Rep* 2001; 8:115-8.

[21] Silva JM, Dominguez G, Garcia JM, Gonzalez R, Villanueva MJ, Navarro F, Provencio M, San Martin S, Espana P, Bonilla F. Presence of tumor DNA in plasma of breast cancer patients: clinicopathological correlations. *Cancer Res* 1999; 59:3251-6.

[22] Ng EK, Tsui NB, Lam NY, Chiu RW, Yu SC, Wong SC, et al. Presence of filterable and non filterable mRNA in the plasma of cancer patients and healthy individuals. *Clin Chem* 2002; 48:1212-7.

[23] Tsui NB, Ng EK, Lo YM. Stability of Endogeneous and added RNA in blood specimens, serum, and plasma. *Clin Chem* 2002; 48:1647-53.

[24] Tatsuma T, Goto S, Kitano S, Lin YC, Lee CM, Chen CL. Telomerase activity in peripheral blood for diagnosis of hepatoma. *J Gastroenterol Hepatol* 2000; 15:1064-70.

[25] Kopreski MS, Benko FA, Kwak LW, Gocke CD. Detection of tumor messenger RNA in the serum of patients with malignant melanoma. *Clin Cancer Res* 1999; 5:1961-5.

[26] Chen XQ, Bonnefoi H, Pelte M-F, Lyautey J, Lederry C, Movarekhi S, et al. Telomerase RNA as a detection marker in the serum of breast cancer patients. *Clin Cancer Res* 2000; 6:3823-6.

[27] Miura N, Shiota G, Nakagawa T, Sano A, Marumoto A, Kishimoto Y, et al. Sensitive detection of hTERT mRNA in the serum of patients with hepatocellular carcinoma. *Oncology* 2003; 64:430-4.

[28] Miura N, Horikawa I, Nishimoto A, Ohmura H, Ito H, Hirohashi S, et al. Progressive telomere shortening and telomerase reactivation during hepatocellular carcinogenesis. *Cancer Genet Cytogenet* 1997; 93:56-62.

[29] Cavin LG, Romieu-Mourez R, Panta GR, Sun J, Factor VM, Thorgeirsson SS, et al. Inhibition of CK2 activity by TGF-beta 1 promotes IkappaB-alpha protein stabilization and apoptosis of immortalized hepatocytes. *Hepatology* 2003; 38:1540-51.

[30] Miura N, Maeda Y, Kanbe T, Yazama H, Takeda Y, Sato R, Tsukamoto T, Sato E, Marumoto A, Harada T, Sano A, Kishimoto Y, Hirooka Y, Murawaki Y, Hasegawa J, Shiota G. Serum human telomerase reverse transcriptase messenger RNA as a novel tumor marker for hepatocellular carcinoma. *Clin Cancer Res* 2005 11:3205-9.

[31] Peng SY, Chen WJ, Lai PL, Jeng YM, Sheu JC, Hsu HC. High alpha-fetoprotein level correlates with high stage, early recurrence and poor prognosis of hepatocellular carcinoma: significance of hepatitis virus infection, age, p53 and beta-catenin mutations. *Intl J Cancer* 2004; 112:44-50.

[32] Takahashi S, Kitamoto M, Takaishi H, Aikata H, Kawakami Y, Nakanishi T, et al. Expression of telomerase component genes in hepatocellular carcinoma. *Eur J Cancer* 2000; 36:496-502.

[33] Waguri N, Suda T, Nomoto M, Kawai H, Mita Y, Kuroiwa T, et al. Sensitive and specific detection of circulating cancer cells in patients with hepatocellular carcinoma; detection of human telomerase reverse transcriptase messenger RNA after immunomagnetic separation. *Cancer Res* 2003; 9:3004-11.

[34] Fujita Y, Fujikane T, Fujiuchi S, Nishigaki Y, Yamazaki Y, Nagase A, et al. The diagnostic and prognostic relevance of human telomerase reverse transcriptase mRNA expression detected in situ in patients with non small cell lung carcinoma. *Cancer* 2003; 98:1008-13.

In: New Messenger RNA Research Communications
Editor: Lee B. Kwang, pp. 99-111
ISBN 978-1-60021-488-2

Chapter VII

Inhibition of RNA Synthesis by 5-Fluorouridine Accounts for its Cytotoxicity on Colorectal Cancer Cells *in Vitro*

Pochi R Subbarayan[6, 1]***, Paul Braunschweiger***[2] ***and Bach Ardalan***[1]
Department of Medicine, [1]Division of Hematology and Oncology, [2]Division of Radiation Oncology, University of Miami, Miller School of Medicine, 1550 NW 10th Avenue, Fox 431A, Miami, FL – 33136, USA

Abstract

Pyrimidine analogue 5-fluorouracil (5-FU) is a standard chemotherapeutic agent in treating colorectal cancer (CRC). Occasionally fluorodeoxyuridine (FUdR) has been used; however, the third analogue fluoro uridine (FUR) has never been used in the clinic. In this study we compared the anti-tumorigenic activity of the above agents *in vitro* and *in vivo*. The cultured cells were protected from the cytotoxic effects of the FUdR by deoxy thymidine and FUR by uridine. Cytotoxicity of these agents as analyzed *in vitro* by cell culture demonstrated FUR to be the most potent anti-tumor agent. *In vivo* we tested the effect of 5-FU and FUR on tumor bearing mice. Intra-peritoneal administration of 5-FU or FUR proved that FUR is fifteen times more effective than 5-FU. Severe inhibitory effect of FUR on TS expression and its RNA mediated cell kill may explain its superiority over the other two fluoropyrimidines tested *in vivo*.

Key words: Colorectal cancer, Fluoropyrimidines, Thymidylate synthase, 5-FU, FUdR, FUR, Animal experiments.

[6] *Corresponding author:Tel.: +1-305-243-4887 E.mail: *spochi@med.miami.edu*

Introduction

The late Dr. Charles Heidelberger in the early 1950's discovered three fluoropyrimidines namely 5-fluorouracil (5-FU), 5-fluoro-deoxyuridine (FUdR) and 5-fluorouridine (FUR)[10]. Since then, 5-FU remains the most commonly used anti-neoplastic agent in colorectal cancer (CRC). The other two fluoropyrimidines were not used because of synthesis problems and inadequate preclinical and clinical studies.

5-fluororouracil is the main agent used in treatment of CRC. 5-fluorodeoxy uridine is approved for liver directed chemotherapy in patients with liver disease only [1,11,20,25,26]. Currently, 5-FU is used in combination with other anti-neoplastic drugs like CPT-11 or Oxaliplatin to treat patients with colon cancer. 5-FU and FUdR are metabolized to 5-fluoro-2'-deoxyuridine monophosphate (FdUMP). FdUMP competes with dUMP to be the substrate for thymidylate synthase (TS), an important enzyme in the formation of dTMP. Incorporation of fluorinated analogues into DNA causes replication defect resulting in cell death. The lesser-known fluoropyrimidine analogue is 5-fluorouridine (FUR). It is incorporated into RNA thereby affecting translation [20,22,30]. Compared to 5-FU and FUdR, FUR is the most cytotoxic of all the fluoropyrimidine analogues [15].

The mode of anti-tumor activity of 5-FU, FUdR and FUR is not completely understood. *In vitro* studies suggest that concentration and duration of exposure determines the cytotoxicity of 5-FU [3,4,8,9,17,21,24]. Clinical studies demonstrate that administration by 24 hrs infusion of 5-FU and FUdR is superior to bolus administration [1,2,16,28,29]. Moreover, 24 hrs high-dose administration of FUdR has better efficacy than 5-FU [1]. Although FUR is the most efficient among the fluoropyrimidine analogues, little information is available regarding its clinical anti-proliferative properties.

Based on the mode of administration, we tried to compare the efficacy of the three fluoropyrimidines *in vitro* and *in vivo*. In this manuscript we present the results of the investigation into the mechanisms of action of 5-FU, FUdR and FUR in *in vitro* cell culture using colorectal cancer cell lines Colo320HSR, CCL221 and HT29. To unravel the mode of action of these drugs we rescued the treated cells with deoxythymidine (dThy) or uridine (Urd). We established time and dose dependent cell kill for fluoropyrimidine analogues on Colo320HSR cells. To determine the effect on transcription, we examined the TS mRNA level in Colo320HSR cells following drug treatment. We also induced tumors in Balb/C mice and challenged with 5-FU or FUR. The anti-tumor activity of 5-FU and FUR was analyzed by clonogenic cell survival assay.

Materials and Methods

Cell Lines and Culture Conditions

The human adenocarcinoma cell lines Colo320HSR, CCL221 and HT29 were obtained from American Type Culture Collection, USA. All the cell lines were maintained in RPMI 1640 medium supplemented with 10% FBS (Invitrogen, USA) at 37°C in humidified 5% CO_2 atmosphere.

Chemicals, Primers and Enzymes

Cell culture medium and supplements were purchased from Invitrogen, USA. 5-FU, FUdR, FUR, dThymidine and Uridine were obtained from Sigma Chemical Company, USA.

Total RNA was prepared using Tri-Reagent (Sigma, USA). Oligonucleotides for cDNA synthesis and PCR were prepared in the University of Miami Miller School of Medicine DNA core facility. The list of primers used and their respective sequences are given in table 1. M-MLV reverse transcriptase and dNTP mix were purchased from Promega, USA. RNase inhibitor was obtained from Applied Biosystem, USA. All other chemicals were reagent grade.

Table 1. List of primers used for PCR.

Primer	Sequence (5' – 3')	Product Length
TS-F	GGG CAG ATC CAA CAC ATC CTC CGC T	970 bp
TS-R	AGG TAA CTG ATA GGT CAC GGA CAG	
β-Actin F	TGA CGG GGT CAC CCA CAC TGT GCC CAT CAT	661 bp
β-Actin R	CTA GAA GCA TTT GCG GTG GAC GAT GGA GGG	

Cytotoxic and Rescue Assays

1 X 10^5 exponentially growing cells were plated in six well plates. Twenty-four hours later various concentrations of 5-FU, FUdR or FUR were added in 2 ml medium and incubated at standard incubation conditions. Rescue studies were performed by simultaneous addition of 10 μM of deoxythymidine (dThy) or 1 mM of uridine (Urd) to the culture. Seventy-two hours later numbers of surviving cells were determined using Coulter Counter (Coulter Electronics, USA) and IC50 values were calculated. Student T test was used to determine degree of significance in IC50 values following dThy or Urd mediated rescue.

Time and Dose Dependent Chemo Sensitivity Studies

To determine the effects of time and concentration of drugs, 1 X 10^5 exponentially growing Colo320HSR cells were exposed to 1, 5 or 10 μM of 5-FU, FUdR or FUR for 1, 5 or 24 hrs at standard culture conditions. At the end of incubation the drug was washed off and the cells were allowed to recover. The viable number of cells over a period of time was determined using Coulter Counter. All the above experiments were carried out in duplicate and repeated four times.

Reverse Transcriptase PCR for TS and β-Actin mRNAs First Strand cDNA Synthesis

One μg of total RNA was reverse transcribed using M-MLV reverse transcriptase. Each reaction mix contained 0.5 μg of oligo dT_{16} primer per μg of the total RNA used. It was heated to 70°C for 5 min and chilled immediately on ice. The final reaction volume of 25 μl contained 0.5 mM each of dNTP mix, 25 U of RNase inhibitor and 200 U of M-MLV reverse transcriptase. The reaction mixture was incubated at 42°C for 60 minutes and stored on ice.

Second Strand Synthesis and Quantitation of TS mRNA PCR

Second strand synthesis was carried out with Gene Amp PCR kit according to the supplier's instructions. The reaction mixture contained 200 nM each of the gene specific primers (table 1) and 5 μl of the first strand cocktail. β-Actin gene was used as an internal reference. The cycling conditions were 30 sec. denaturation at 94°C, 30 sec. annealing at 60°C and 60 sec. extension at 72°C for 30 cycles. The reaction products were analyzed by electrophoresis through 1.0 % agarose gel and photographed.

In Vivo Effect f 5-FU and FUR

Mouse colorectal cancer cell line CT24.WT [27] was purchased from ATCC, USA. It was maintained in RPMI 1460 medium supplemented with 10% FBS and 1% PS. Five week old Balb/C mice (Harlan Sprague Dawley, USA) were injected with approximately 3 x 10^5 CT26.WT cells in 100-200 μl of PBS. Three weeks later, when tumors were 1.5 cm^2 in size, 2.5 mg FUR or 12.5 mg of 5-FU per kg body weight were injected IP to the tumor bearing mice. Twenty four hours later, the animals were killed in a CO_2 chamber. The tumors were extracted under sterile conditions, rinsed two times in sterile PBS, minced and transferred to sterile ice cold trypsin-EDTA. Following over night incubation in cold trypsin single cell suspension was obtained. Clonogenic assays were performed by plating 50 or 100 cells in duplicate wells of 24 well plates. Two weeks later, formed colonies were counted and recorded. Each experimental set contained at least three animals and were repeated three times. Animal experiments were carried out as per ACUC approved protocol No. 05-070.

RESULTS

FUR and FUDR are 150 Fold More Effective Than 5-FU

Our aim is to compare the antitumorigenic potential of the three uridine analogues. We used cytotoxicity assays to compare their efficacy as chemotherapy agents. The results are presented in Table 2. The data show that nanomolar quantities of FUdR and FUR are sufficient to kill 50% of cells compared to micromolar concentrations of 5-FU. This is in agreement with a previous report [15]. There appears to be cell line dependent variation in the drug dosage. Some cell lines may be intrinsically more resistant to anti-tumor agents. In particular CCL221 is about 1.4 to 2.7 folds more resistant to all the three drugs compared to Colo320HSR and HT29. In contrast the latter two cell lines exhibited almost similar IC50 values (table 2).

Table 2. IC 50 values (μM) of 5-FU, FUdR and FUR for Colo320HSR, CCL221 and HT29 cells.

	No rescue	+ 10 μM dThy	*P-Value*	+1 mM Urd	P-Value
5-FU					
Colo320HSR	3.278 + 0.467	5.91 + 0.857	0.041	3.338 + 0.453	0.175
CCL221	8.732 + 1.348	30.710 + 9.861	0.017	18.521+4.221	0.029
HT29	6.062 + 0.536	7.200 + 1.084	0.348	3.548 + 1.549	0.135
FUdR					
Colo320HSR	0.019 + 0.007	194.062 + 63.982	0.0005	0.017 + 0.003	0.781
CCL221	0.029 + 0.007	671.272 + 84.578	0.00000005	0.028 + 0.011	0.0927
HT29	0.014 + 0.004	155.621 + 50.600	0.00039	0.024 + 0.006	0.247
FUR					
Colo320HSR	0.023 + 0.007	0.02 + 0.005	0.738	2.395 + 0.468	0.0001
CCL221	0.052 + 0.009	0.056 + 0.006	0.786	6.861 + 1.419	0.0000073
HT29	0.041 + 0.003	0.059 + 0.010	0.140	2.700 + 0.685	0.008

The IC50 values were determined as described under materials and methods. The values are Mean ± SE from at least four independent experiments done in duplicate. P value indicates the degree of significance between rescue and cytotoxic studies.

Deoxythymidine Protected the Cells from FUDR Whereas Uridine from the Effects Of FUR

10 μM deoxythymidine (dThy) uniformly rescued all three cell lines from the cytotoxic effects of FUdR, but not from FUR. In contrast 1 mM uridine (Urd) was required to protect the cells from the toxic effects of FUR (table 2).

FUR Exhibits Superior Cytotoxicity at Bolus and Infusion Doses

We simulated bolus administration of drugs by short term exposure (1 h) of Colo320HSR cells in culture whereas 5 and 24 hrs replicated infusional regimen. The results of time and dose dependent effect of 5-FU, FUdR and FUR on Colo320HSR are presented in thefigure 1 A through I.

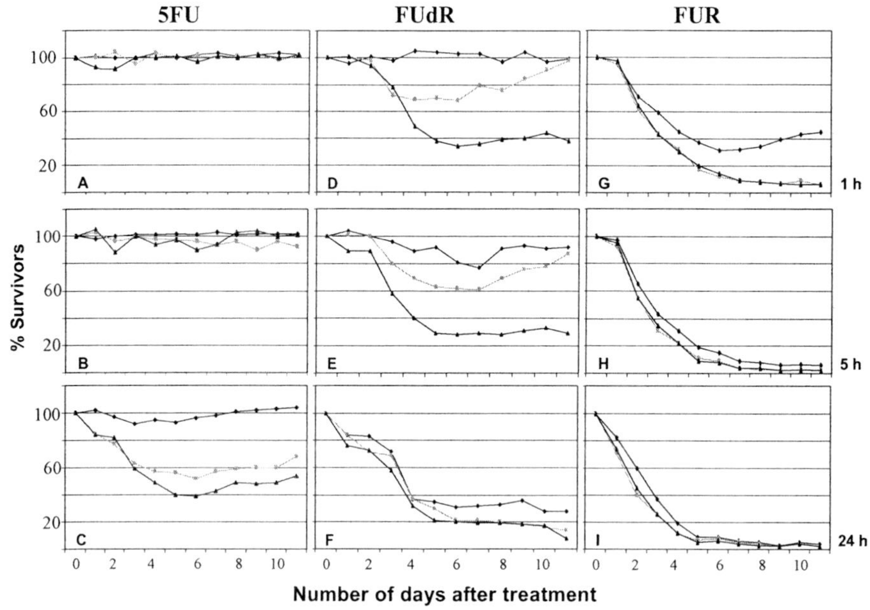

Figure 1. Time and dosage dependent cytotoxic effects of fluoropyrimidines on Colo320HSR cells. FUR at all the conditions tested was found to be cytotoxic. 1 X 10^5 exponentially growing Colo320HSR cells were exposed to 1 μM (closed diamonds), 5 μM (closed squares) or 10 μM (closed triangles) of 5-FU, FUdR or FUR for 1 h (panels A,D,G), 5 h (panels B,E,H), or 24 h (panels C,F,I) at standard incubation conditions. At the end of the incubation period the drug was washed off and the cells were allowed to recover. The viability over a period of time was determined by tryphan blue exclusion assay using Coulter Counter (Coulter Electronics, FL). These experiments were carried out in duplicate and repeated four times.

5-FU is Most Effective Only at High Dose and Longer Treatment Time

Short time (1 h) exposure of Colo320HSR cells to 5-FU at low (1 μM) to moderate dose (5 mM) did not induce cell death (figure 1A, B). Lower dosage (1 μM) in combination with longer duration (24 h; figure 1C), moderate dosage (5 μM) for intermediate duration (5 h; figure 1B) and high dosage (10 μM) for a short time (1 h; figure 1A) were moderately (2-10%) cytotoxic. A moderate to high dose (5, 10 μM) of 5-FU for a longer time (24 h) inhibited growth by 40-60% (figure 1C).

High Dose of FUDR is Cytotoxic as Bolus and Infusion Regimen

Treatment of Colo320HSR cells for 1 h with 1 μM FUdR was least cytotoxic (figure 1 D), whereas treatment for 5 h killed about 20% of cells (figure 1 E). A moderate dose of 5 μM for 1 or 5 h was effective in inhibiting cell growth by as much as 40% (figure 1D, E). Exposure of cells to a low dose (1 μM) for 24 h (figure 1F), high dose (10 μM) for 1 h (figure 1D) or 5 h (figure. 1E) inhibited growth of 60–70% of cells. Longer treatment period (24 h) at all the concentrations (1, 5, 10 μM) killed most of the cells; i.e. 70-90% growth inhibition (figure. 1 F).

High or Low, Bolus or Infusion of FUR Kills the Most

A combination of low dose (1 μM) of FUR and short exposure time caused 60% growth inhibition (figure 1 G). All other combinations viz, low dose (1 μM) with moderate (5 h; figure 1H) to long (24 h; figure 1I) exposure time, or moderate to higher dose for short (figure 1G), moderate (figure 1H) or long duration (figure 1I) killed most of the cells (80-95% growth inhibition).

FUR Suppressed TS Gene Expression in Cultured Cells Under Short and Long Term Exposure

It is suggested that modified uridine derivative (FUTP) is incorporated into messenger RNA, thus affecting its stability and function. Therefore, we checked the effect of FUR on mRNA synthesis in Colo320HSR cells. 5-FU and FUdR inhibit the activity of the enzyme thymidylate synthase (TS) at protein level. We evaluated the effect of all the three drugs on mRNA synthesis by quantitating TS transcripts using semi-quantitative reverse transcriptase PCR methodology. The house keeping gene β-actin mRNA was used as a control. Results of this experiment are presented in Figures 2 and 3. Two hrs treatment of Colo320HSR cells with 0.1 μM concentration of fluoropyrimidines increased TS expression (figure. 2A lanes 2, 3, 4); whereas, a longer duration of 24 hrs decreased TS transcription in FUdR and FUR treated samples (figure 2 B lanes 3, 4). Exposure of these cells to 1 μM of 5-FU, FUdR or FUR whether for 2 or 24 hrs uniformly inhibited the expression of TS (figure 3 A, B lanes 2, 3, 4). We also noted that treating cells for 24 hrs with 0.1 μM of FUdR or FUR was more potent in inhibiting TS transcription than 5-FU. One μM of FUR in 2 or 24 h inhibited TS gene expression more than either 5-FU or FUdR (figure 3 A, B lane 4). In the same samples, mRNA quantity of the house-keeping gene β-actin was not affected (figures 2C, D, 3C, D)

FUR is 15 Fold More Effective than 5-FU in its *In Vivo* Anti-Tumor Activity

Having checked the efficacy of these three agents on cultured cells, we tested the effects of 5-FU and FUR on tumor bearing Balb/C mice.figure 4 summarizes the results of this study. Under the same treatment conditions 2.5 mg/kg of FUR killed 78% of cells compared to 26% by 12.5 mg/Kg 5-FU. It is a difference of 3 fold. Accounting for the difference in the dosage between these two anti-tumor agents; FUR is 15 times more potent than 5-FU *in vivo*, which is comparable to the *in vitro* data.

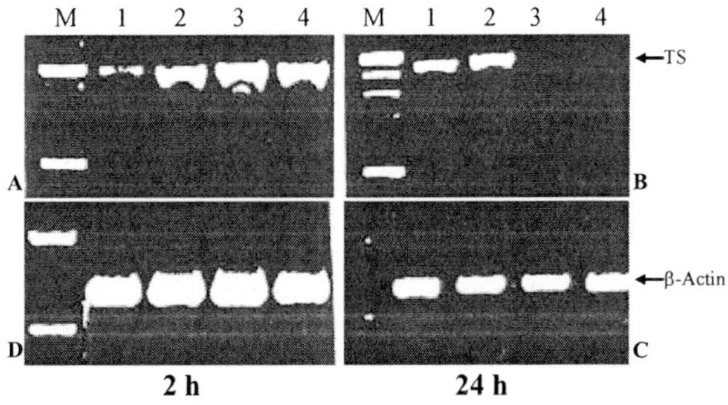

Figure 2. Agarose gel analysis of TS and β-actin RT-PCR products. 0.1 μM of FUdR and FUR suppressed TS gene expression in Colo320HSR cells. Total RNA was isolated from treated cells, cDNA synthesis and PCR for TS and β-actin genes were carried out using gene specific primers (table 1) as described under materials and methods. PCR products were analyzed on 1% agarose gel, stained with ethidium bromide and documented. Lane M. DNA Marker. Lane 1. Untreated control. Lane 2. 5-FU, Lane 3. FUdR, Lane 4. FUR. Panel A and D 2 hrs exposure, Panel B and C 24 hrs exposure.

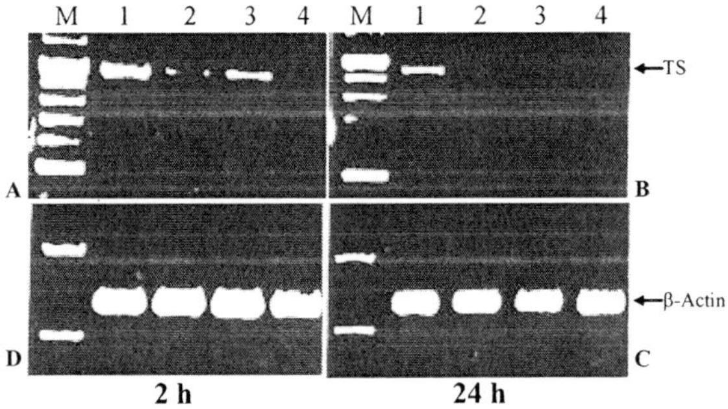

Figure 3. Agarose gel analysis of TS and β-actin RT-PCR products. 1 μM of 5-FU, FUdR and FUR suppressed TS gene expression in Colo320HSR cells. Details of the experiment are as described under Materials and Methods. Lane M. DNA Marker. Lane 1. Untreated control. Lane 2. 5-FU, Lane 3. FUdR, Lane 4. FUR. Panel A and D 2 hrs exposure, Panel B and C 24 hrs exposure.

Discussion

5-FU has been the drug of choice in treating many solid tumors. Its analogue FUdR is only approved for liver directed therapy[1]. Due to its potential toxicity FUR had never been put into clinical trial. In this study we compared the *in vitro* and *in vivo* cytotoxicity of FUR with 5-FU. In the *in vitro* cell culture one hour treatment replicated bolus administration, while 5 and 24 hrs reflected infusion regimen of 5-FU, FUdR and FUR.

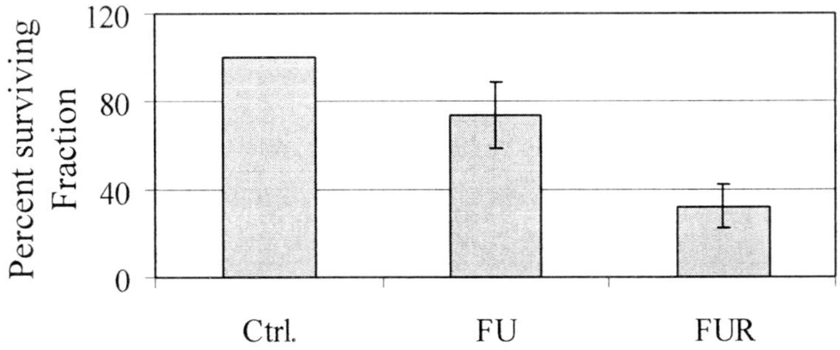

Figure 4. Flurouridine is 15 fold more toxic than 5-FU in its antitumorigenic activity on tumors induced in Balb/C mice. Mouse colorectal cancer cell line CT26.WT was injected S.C. to Balb/C mice. Following tumor formation mice were treated with PBS, 5-FU or FUR. 24 hrs later, tumors were extracted and clonogenic assay performed. At least three animals constituted a group and each experiment was performed three times.

As represented in Figure 1 in general, infusion of these drugs is more effective than bolus administration. Previous studies showed an inverse relationship between the duration of exposure and concentration of 5-FU used. High dose (>100 μM) for a short period of time (<6h) or 1 – 10 μM for longer than 72 hrs was found to be effective on a variety of tumor types [3,4,8,9,13,16-18,21,24]. 5-FU is rapidly cleared from the plasma; therefore, infusion ensures a steady supply which leads to effective cytotoxicity. In short 5-FU infusion seems be significantly effective and is in agreement with clinical observations [1,13,16,18,28].

We note that cells were able to recover from the damaging effects of short to moderate exposure to FUdR (figure 1E). However, even low dose (1 μM FUdR) treatment for longer time eliminated this ability (figure 1F). This reinforces the notion that infusion of FUdR is superior to bolus administration. Tumor tissues are reported to be deficient in phosphate donors. Inside the malignant cells FUdR is predominantly converted to 5-FU and eventually to FUMP. 5-FU appears to be more stable in tissue than the extra cellular fluids[12]. Long term exposure to FUdR at all concentrations may lead to the accumulation of more metabolites inside the cultured cells causing maximal damage accounting for the efficacy of FUdR. In contrast, a high dose of FUdR killed 60 – 80% of cells at all the treatment time points. However, 5 and 10 μM of FUdR is not a clinically achievable concentration.

Cytotoxicity and animal data demonstrate 5-fluorouridine (FUR) as a potent anti-tumor agent. Irrespective of the dosage and duration of exposure, FUR killed 60-90% of cells. The cytotoxicity of FUR is rescued by uridine. We also observed that mRNA synthesis is affected by FUR. This data is consistent with the hypothesis that FUR interferes with RNA pathway. Incorporation of fluoro compounds into RNA is attributed to their toxicity. On the other hand tumor bearing mice received significantly low doses of 5-FU and FUdR and exhibited no toxicity (24 hrs exposure). At a relatively low dose, FUR is 15 folds more effective than 5-FU in killing *in vivo* mouse tumor. Taken together the data validate the superiority of FUR as a chemotherapeutic agent.

There is about 500 fold difference in the IC50 values of 5-FU and FUdR and FUR. It may be due to transportation of 5-FU across the plasma membrane. Variation in transport rate in 5-FU across the plasma membrane had been observed in different cell lines. There is a 17 fold difference in the rate of (^{3}H)-uracil uptake in CaCO-2 and LLC-PK1 cells *in vitro*[14]. The same investigators also observed that sodium enhances uracil transport. Once inside the cell a majority of the 5-FU is catabolized to FDHU by the enzyme dihydro pyrimidine dehydrogenase (DPD). The clinical effectiveness of 5-FU had been correlated to balance among DPD, TP/TP and TS. It is reported that more DPD is a cause for treatment failure or requirement of high dose of 5-FU [23]. We speculate that these factors may play a role in the multifold high requirement of 5-FU compared to FUdR or FUR.

In rescue experiments deoxythymidine could partially inhibit the cytotoxicity of 5-FU (Table 2). The reason for failure of dThy to rescue the cells from the cytotoxic effects of 5-FU could lie with its amount used in competition/rescue experiments. Ten μM of dThy is about three folds more than the observed IC50 values for 5-FU. In contrast 10 μM dThy is a few hundred folds more than the IC50 values of FUdR. Therefore, by increasing the concentration of dThy its inhibitory action on 5-FU could have been enhanced.

Uridine could rescue cells from the damaging effects of FUR only. FUR treatment inhibited TS transcription as noted from semi-quantitative RT-PCR analysis. These facts affirm that FUR mainly affects the RNA pathway causing maximum, immediate and irrevocable damage to cells. Therefore, damage to the functional units namely RNA and thereby protein synthesis triggered by FUR might pose lethal and irrecoverable damage to cells.

The inhibitory effect of 5-FU on TS is correlated with its cytotoxicity. The time and dose dependent cytotoxicity of 5-FU may be due to its short half-life or is limited to proliferating cells [19]. Thus, quiescent cells might escape the effect of 5-FU. However, higher concentration and prolonged exposure of 5-FU may result in additional but not yet understood mechanisms that are independent of cell cycle and TS or incorporation into DNA and RNA. Although FUdR is more effective than 5-FU in killing Colo320HSR, from the data it appears that the inhibitory effects of TS gene expression by 5-FU and FUdR are very similar at short exposure time. But, at longer exposure time, FUdR completely inhibited TS transcription. Greater cytotoxicity of FUdR over 5-FU may be due to its partial conversion to stable 5-FU inside the tumor and effective inhibition of dTMP formation. More FdUMP may be formed following FUdR administration than 5-FU. As a result, more TS accumulates shutting down its transcription [5-7]. It is also possible that formed 5-FU metabolize to FUMP which may eventually be incorporated into RNA forming labile transcripts. Therefore, the mechanism of inhibition of TS transcription following fluorodeoxyuridine administration needs further investigation.

Inhibition of TS transcription by FUR is more severe. FUR is not known to bind TS. Therefore, general perturbance in transcription by FUR may be the cause for the loss of TS transcripts. In the present study, FUR inhibited TS transcription but very little difference in β-actin mRNA level was observed. We think that transcriptional interference occurs gradually. In the absence of cellular proliferation β-actin, a house keeping gene might not be transcribed. As a consequence it might take longer to see any change in its quantity. Thus RNA mediated inhibition of tumor growth by FUR makes it an effective alternate to 5-FU and FUdR.

Toxicity data established the LD 50 values for 5-FU, FUdR and FUR to be about 260 mg/Kg, 1400 mg/kg and 160 mg/kg respectively in CBA/J mice[12] and 450 mg/Kg,

2500mg/kg and 175 mg/Kg for CBA mice (Ardalan et. al. unpublished data) administered in 24 hrs. Accordingly FUR is 2.6 times more toxic than 5-FU and FUdR is least toxic. In contrast 2.5 mg/kg of FUR (about 70 fold less than its LD50) is 15 times more potent than 5-FU (12.5 mg/kg = 36 fold less than its LD50). In conclusion, our studies confirm that among the three fluoropyrimidines tested FUR is the most potent anti-tumor agent. Severe inhibitory effect of FUR on gene expression and its RNA mediated cell kill may explain its superiority over 5-FU and FUdR. In contrast FUR was found to be effective in inhibiting cell growth *in vitro* and *in vivo* at all concentrations and modes of administration. The data also show that FUR at a very low dose (i.e. 70 folds less than its LD50 values in mice) is about 15 folds more cytotoxic than 5-FU. Under appropriate dosage, FUR may become less or equi-toxic but an effective alternate to 5-FU.

ACKNOWLEDGEMENT

We wish to thank Dr. Zeming Deng and Philip Lee for their excellent technical help and other lab members for critical reading of the manuscript.

REFERENCE

[1] Ardalan, B. and Lima, M. (2004) A phase II trial of FUdR in patients with advanced pancreatic cancer. *J.Cancer Res.Clin.Oncol.* 130, 561-566

[2] Ardalan, B., Sridhar, K. S., Benedetto, P., Richman, S., Waldman, S., Morrell, L., Feun, L., Savaraj, N., Fodor, M., and Livingstone, A. (1991) A phase I, II study of high-dose 5-fluorouracil and high-dose leucovorin with low-dose phosphonacetyl-L-aspartic acid in patients with advanced malignancies. *Cancer* 68, 1242-1246

[3] Aschele, C., Sobrero, A., Faderan, M. A., and Bertino, J. R. (1992) Novel mechanism(s) of resistance to 5-fluorouracil in human colon cancer (HCT-8) sublines following exposure to two different clinically relevant dose schedules. *Cancer Res.* 52, 1855-1864

[4] Calabro-Jones, P. M., Byfield, J. E., Ward, J. F., and Sharp, T. R. (1982) Time-dose relationships for 5-fluorouracil cytotoxicity against human epithelial cancer cells in vitro. *Cancer Res.* 42, 4413-4420

[5] Chu, E., Koeller, D. M., Johnston, P. G., Zinn, S., and Allegra, C. J. (1993) Regulation of thymidylate synthase in human colon cancer cells treated with 5-fluorouracil and interferon-gamma. *Mol.Pharmacol.* 43, 527-533

[6] Chu, E., Voeller, D., Koeller, D. M., Drake, J. C., Takimoto, C. H., Maley, G. F., Maley, F., and Allegra, C. J. (1993) Identification of an RNA binding site for human thymidylate synthase. *Proc.Natl.Acad.Sci.U.S.A* 90, 517-521

[7] Codacci-Pisanelli, G., van der Wilt, C. L., Pinedo, H. M., Franchi, F., Noordhuis, P., Braakhuis, B. J., van Laar, J. A., and Peters, G. J. (1995) Antitumour activity, toxicity and inhibition of thymidylate synthase of prolonged administration of 5-fluorouracil in mice. *Eur.J.Cancer* 31A, 1517-1525

[8] Drewinko, B. and Yang, L. Y. (1985) Cellular basis for the inefficacy of 5-FU in human colon carcinoma. *Cancer Treat.Rep.* 69, 1391-1398

[9] Evans, R. M., Laskin, J. D., and Hakala, M. T. (1980) Assessment of growth-limiting events caused by 5-fluorouracil in mouse cells and in human cells. *Cancer Res.* 40, 4113-4122

[10] Heidelberger, C. (1965) Fluorinated pyrimidines. Prog.Nucleic Acid *Res.Mol.Biol.* 4, 1-50

[11] Heidelberger, C., Danenberg, P. V., and Moran, R. G. (1983) Fluorinated pyrimidines and their nucleosides. *Adv.Enzymol.Relat Areas Mol.Biol.* 54, 58-119

[12] Houghton, J. A., Houghton, P. J., and Wooten, R. S. (1979) Mechanism of induction of gastrointestinal toxicity in the mouse by 5-fluorouracil, 5-fluorouridine, and 5-fluoro-2'-deoxyuridine. *Cancer Res.* 39, 2406-2413

[13] Kish, J. A., Ensley, J. F., Jacobs, J., Weaver, A., Cummings, G., and Al Sarraf, M. (1985) A randomized trial of cisplatin (CACP) + 5-fluorouracil (5-FU) infusion and CACP + 5-FU bolus for recurrent and advanced squamous cell carcinoma of the head and neck. *Cancer* 56, 2740-2744

[14] Li, H., Chung, S. J., and Shim, C. K. (2002) Characterization of the transport of uracil across Caco-2 and LLC-PK1 cell monolayers. *Pharm.Res.* 19, 1495-1501

[15] Link, K. H., Aigner, K. R., Peschau, K., Warthona, M., Schwemmle, K., and Danenberg, P. V. (1988) Concentration and time dependence of the toxicity of fluorinated pyrimidines to HT 29 colorectal carcinoma cells. *Cancer Chemother.Pharmacol.* 22, 58-62

[16] Lokich, J. J., Ahlgren, J. D., Gullo, J. J., Philips, J. A., and Fryer, J. G. (1989) A prospective randomized comparison of continuous infusion fluorouracil with a conventional bolus schedule in metastatic colorectal carcinoma: a Mid-Atlantic Oncology Program Study. *J.Clin.Oncol.* 7, 425-432

[17] Moran, R. G. and Scanlon, K. L. (1991) Schedule-dependent enhancement of the cytotoxicity of fluoropyrimidines to human carcinoma cells in the presence of folinic acid. *Cancer Res.* 51, 4618-4623

[18] O'Connell, M. J., Martenson, J. A., Wieand, H. S., Krook, J. E., Macdonald, J. S., Haller, D. G., Mayer, R. J., Gunderson, L. L., and Rich, T. A. (1994) Improving adjuvant therapy for rectal cancer by combining protracted-infusion fluorouracil with radiation therapy after curative surgery. *N.Engl.J.Med.* 331, 502-507

[19] Pestalozzi, B. C., McGinn, C. J., Kinsella, T. J., Drake, J. C., Glennon, M. C., Allegra, C. J., and Johnston, P. G. (1995) Increased thymidylate synthase protein levels are principally associated with proliferation but not cell cycle phase in asynchronous human cancer cells. *Br.J.Cancer* 71, 1151-1157

[20] Pinedo, H. M. and Peters, G. F. (1988) Fluorouracil: biochemistry and pharmacology. *J.Clin.Oncol.* 6, 1653-1664

[21] Ren, Q., van Groeningen, C. J., Hardcastle, A., Aherne, G. W., Geoffroy, F., Allegra, C. J., Johnston, P. G., and Grem, J. L. (1997) Determinants of cytotoxicity with prolonged exposure to fluorouracil in human colon cancer cells. *Oncol.Res.* 9, 77-88

[22] Rustum, Y. M. (1991) Modulation of fluoropyrimidines by leucovorin: rationale and status. *J.Surg.Oncol.Suppl* 2, 116-123

[23] Salonga, D., Danenberg, K. D., Johnson, M., Metzger, R., Groshen, S., Tsao-Wei, D. D., Lenz, H. J., Leichman, C. G., Leichman, L., Diasio, R. B., and Danenberg, P. V. (2000) Colorectal tumors responding to 5-fluorouracil have low gene expression levels of dihydropyrimidine dehydrogenase, thymidylate synthase, and thymidine phosphorylase. *Clin.Cancer Res.* 6, 1322-1327

[24] Santelli, G. and Valeriote, F. (1986) Schedule-dependent cytotoxicity of 5-fluorouracil in mice. *J.Natl.Cancer Inst.* 76, 159-164

[25] Thomas, D. M. and Zalcberg, J. R. (1998) 5-fluorouracil: a pharmacological paradigm in the use of cytotoxics. *Clin.Exp.Pharmacol.Physiol* 25, 887-895

[26] van Laar, J. A., Rustum, Y. M., Ackland, S. P., van Groeningen, C. J., and Peters, G. J. (1998) Comparison of 5-fluoro-2'-deoxyuridine with 5-fluorouracil and their role in the treatment of colorectal cancer. *Eur.J.Cancer* 34, 296-306

[27] Wang, M., Bronte, V., Chen, P. W., Gritz, L., Panicali, D., Rosenberg, S. A., and Restifo, N. P. (1995) Active immunotherapy of cancer with a nonreplicating recombinant fowlpox virus encoding a model tumor-associated antigen. *J.Immunol.* 154, 4685-4692

[28] Webb, A., Cunningham, D., Scarffe, J. H., Harper, P., Norman, A., Joffe, J. K., Hughes, M., Mansi, J., Findlay, M., Hill, A., Oates, J., Nicolson, M., Hickish, T., O'Brien, M., Iveson, T., Watson, M., Underhill, C., Wardley, A., and Meehan, M. (1997) Randomized trial comparing epirubicin, cisplatin, and fluorouracil versus fluorouracil, doxorubicin, and methotrexate in advanced esophagogastric cancer. *J.Clin.Oncol.* 15, 261-267

[29] Whitehead, R. P., Benedetti, J. K., Abbruzzese, J. L., Ardalan, B., Williamson, S., Gaynor, E. R., Balcerzak, S. P., and Macdonald, J. S. (2004) A phase II pilot study of high-dose 24-hour continuous infusion of 5-FU and leucovorin and low-dose PALA for patients with colorectal cancer: a Southwest Oncology Group study. *Invest New Drugs* 22, 467-473

[30] Zhang, Z. G., Harstrick, A., and Rustum, Y. M. (1992) Modulation of fluoropyrimidines: role of dose and schedule of leucovorin administration. *Semin.Oncol.* 19, 10-15

In: New Messenger RNA Research Communications
Editor: Lee B. Kwang, pp. 113-119
ISBN 978-1-60021-488-2

Chapter VIII

Passive Repetitive Stretches Activate the Skeletal Muscle and Induce the Expression of mRNA of Transcription Factors

S. Ikeda, Y. Kamikawa, K. Harada, A. Oowatashi, A.Yoshida, K. Horinouchi and K. Kawahira

Department of Rehabilitation and Physical Medicine, Graduate School of Medical and Dental Sciences, Kagoshima University. Japan

Abstract

Skeletal muscle is the biggest organ in the human body and its role is to move the body. Muscle growth and strengthening is important theme on sports or clinical medicine. Mechanical stretch has been considered to be a growth stimulation for muscle cells. Continuous stretching increased muscle weight in vivo and repetitive stretching were also reported to induce cellular growth for cultured cardiac muscle cells. We have investigated the effect of repetitive passive stretch to muscle tissues using molecular biological methods. Passive repetitive stretching induce the expression of immediate early gene, myogenic transcription factor to growth the skeletal muscles. Recent studies clarify the mechano-transduction and mechanisms about muscle growth with gene expressions. Mechanical stretching to skeletal muscle stimulates the mechanoreceptors on the surface of muscle cells and activates the intracellular signal pathway to gene activation. Then, gene expression of immediate early gene, myogenic transcription factor, growth factor, muscle structure gene were occurred and muscle growth were obtained. We introduce mechanical stimulation induce the muscle growth with expressions of mRNAs.

INTRODUCTION

Muscle strengthen exercise is oldest theme in rehabilitation medicine and sports science. Furthermore, one of the most important problems now. Muscle training increase the muscle power on the athlete as well as improves moving activity on the patients of neurological disorder, musculoskeletal disease or disuse muscle atrophy by bed ridden. Muscle strengthening improves the sports performance and activity of daily living or quality of life. Although gain of muscle power is considered that the muscle hypertrophy or growth and efficient of neurotransmission, muscle hypertrophy is very important for muscle strengthening. Recently, mechanical stimulation, stretching, is considered to have a grate role on muscle hypertrophy. The mechanism of stretch-induced hypertrophy gradually becomes to know by the molecular biological technique.

MUSCLE STRENGTHEN EXERCISE FROM THE VIEW POINT OF MOLECULAR BIOLOGY

Skeletal muscle is composed of striated muscle fibers and muscle contraction is the results of sliding action between myosin filament and actin filament induced by chemical reaction on nervous transmission. Because of muscle power is correlated with muscular sectional area [1], muscle strengthen training has close relationship with muscle hypertrophy. Muscle hypertrophy is the results of increased protein synthesis of muscle structure elements etc. Myosin, actin, troponin, or enzymes of muscular metabolism, etc. followed by gene expression. These proteins are the polypeptides composed of linkage of amino acid. These sequence information are coded in DNA and transcripted into mRNA, then translated to proteins that have various functions. From the viewpoint of molecular biology, we can consider that muscle strengthen training is the results of gene activation and transcription, translation.

STRETCH AND MUSCLE HYPERTROPHY

Various muscle exercise aim to increase muscle power are done. Although there are systemic effects on muscle exercise, major effects are occurred on the trained muscle itself. One of the explanations of focal muscle hypertrophy is tropic factor from nerve. Recently, mechano-receptors in the muscle cell were become to know, stretching is considered to be important factor to muscle hypertrophy. Vandenburgh et. al. described that hypertrophy were seen in the cultured muscle cell on the stretchable silicone tray at stretch stimulation [2]. These findings indicate that stretch stimulation induced the muscle hypertrophy. The study in vivo, stretch overload of the avian latissimus dorsi muscle increased muscle mass [3] with muscle fiber formation [4] and fiber hypertrophy [5]. Furthermore, passive stretching of skeletal muscle induced various gene expressions in vivo and cultured muscle cells. Mechanical stretching for muscle cell enhances DNA transcription and induces the immediate early genes like c-*fos* or c-*jun*, marker of cellular activation [6], then expresses the myogenic transcription factors, myogenin, MyoD, Myf5 [7], finally, gene expression of muscle structure protein like myosin, actin, troponin etc.[8] Following studies clarified the presence of mechano-sensitive receptor on the surface of the muscle cells and transmits into the intra cellular signaling system, then hypertrophy is occurred with protein formation. We reported that passive repetitive stretching of rat removed skeletal muscle for 2 hours induced the expression of immediate early gene c-fos mRNA and increased expression of myogenin on 4hours stretching [9].

Stretch Receptors

In vivo, cells are connected with extracellular matrix each other. Therefore, on a muscle cell occurred contraction, near cell receives the mechanical stimulation. Usual muscular contractions are considered to mechanical stimulation themselves. The receptors to detect the mechanical stimulation on muscle cell have been found recently. These receptors receive the mechanical stimulations and convert into the chemical signals to intracellular cascade, then gene induction and protein production.

Integrin is a dimer protein of stretch receptor. It connects extracellular matrixes and intracellular cytoskeletons being found as an cell adhesion[10]. Later, its function have been clarified, it has an important role of mechanical stimulation. Integrin plays a various role as gene expression, cellular growth, differentiation, apoptosis etc. by mechanical detector. [11,12]

Receptor type tyrosine kinase responsible to growth factors has membrane portion. According to a study about vascular smooth muscle, Platelet derived growth factors (PDGF) receptors are activated intracellular kinase portion by mechanical stimulation and activate the following intracellular signal pathway, then induce gene activation.[13]

There are channels that through Ca2+ or Na+ on mechanical stress found in cultured cardiac muscles.[14] These are called as stretch activated ion channel. These were found in vascular smooth muscles and skeletal muscle, there are non-selective cation channel, ATP sensitive K+channel, Na-H exchange channel etc. These receptors increase the intracellular cations and activate a intracellular signaling system, then facilitate the transcription of DNA and produce the functional proteins [15].

Mechanical Stretch and Growth Factors

Another important factors are growth factors. Growth hormone or other growth factors thought to have relationship. On skeletal muscle Insulin like growth factor (IGF) and Fibroblast growth factor (FGF) is thought to have close relationship with muscle growth and hypertrophy. There have been reported that mechanical stretching induce the increase of IGF or FGF[16,17]

Recently, findings have been reported about local mechanisms of growth factors lead to muscle hypertrophy without systemic humoral effects. [18] According to the study on skeletal muscle, cardiac muscle and smooth muscle, presence of autocrine or paracrine growth factor have been clarified.[19] On mechanical stimulation muscle cell secretes the local growth factor and focal muscular hypertrophy-growth are obtained. Especially, autocrine type insulin like growth factor variant is noted. This growth factor is called Mechano-growth factor as another name, the role is considered to be important for muscle hypertrophy and strengthening.[20] Mechanical stretch stimulation induce the secretion of growth factor, then gene activation were occurred.

Mechanical Stretch to Muscle Growth-Hypertrophy and Gene Expresion

The process of skeletal muscle growth-hypertrophy and increase the muscle strength is summarizing on figure 1.

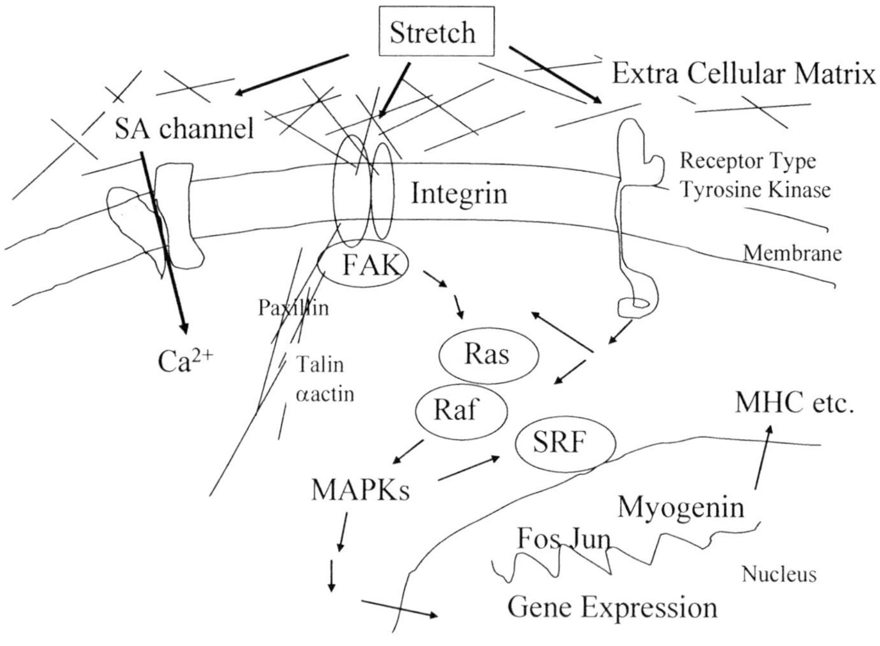

Figure 1. Mechanisms of stretch-induced muscle growth-hypertrophy. Mechanical stretch stimulates the surface mechanoreceptor and intracellular signaling activation. Then gene expression and protein synthesis are accured. FAK: focal adhesion kinases, MAPK: mytogen-activated protein kinases, SRF: Serum Response Factor, SA channel: Stretch activated ionchannel, MHC: myosin heavy chain.

On mechanical stimulation caused by stretching activate the mechanoreceptor like integrin, chemical information of intracellular signaling systems are activated. On the intracellular cascade, integrin activates focal adhesion kinases (FAK), paxillin, mytogen-activated protein kinases (MAPKs). Then transcription of DNA is induced. At first, immediate early genes like c-fos or c-jun mRNA are induced. These products are act on the DNA as transcriptor to induce following gene activation. Next, genes of myogenic transcription factors like myogenin, MyoD, Myf5 etc. are induced. These transcription factors are muscle specific and promote the synthesis of muscle structure protein, myosin, actin, vimentin, troponin etc. Then muscle growth-hypertrophy is occurred and muscle strengthening is obtained. In these processes, autocrine production of MGF thought to accelerate the muscle growth. These gene expression is controlled at transcription procedure, DNA to mRNA, by transcription factor like myogenin or MyoD etc and extracellular information detected by integrin or others affects the expression□[11].

Repetitive Passive Stretch on Skeletal Muscle and Gene Expression

We have been studied the effects on passive repetitive stretching of skeletal muscles. Rat anterior tibial muscle and soleus muscle were removed and repetitively stretched about 20% of the muscle length 15 times per minute using the Magnus method in Krebs Ringer solution. Repetitive stretches were continued for 2 hours in one group and 4 hours in another group. Contralateral muscles removed from the same rat were incubated without stretch as the control. After repetitive stretch, total RNA was extracted from the muscles and northern blotting was carried out using c-*fos* and myogenin oligonucleotide as probes. Repetitive stretch during 2 hours induced a significant increase of c-*fos* mRNA expression in both the tibialis anterior and soleus muscles. Increased expression of myogenin mRNA in both muscles was observed by repetitive stretch for 4 hours. The initial increase of immediate early gene c-*fos* mRNA and the subsequent increase of myogenic transcription factor myogenin mRNA is considered to indicate that sequential progression of gene transcription towards muscle hypertrophy is induced by repetitive muscle stretch. [9]

We introduce another study of passive repetitive stretching of skeletal muscle. Repetitive stretch (15 times / min) was performed manually on gastrocunemius muscle of anesthetized rats for 15, 30, and 60 minutes. Animals were sacrified 24 hours after the manual stretch. Total RNA was then extracted from the gastrocunemius muscle and northern blotting was performed using oligonucleotide complementary to myogenin.

Repetitive stretch gastrocunemius muscles of anesthetized rats for 1 hour induced increased expression of myogenin mRNA by 21.5%. Repetitive passive stretch of the rat skeletal muscle for one hour induced the expression of myogenic transcription factor myogenin mRNA[21].

Mechanical repetitive stretch for a few hours would be one of the stimulating factors for muscle growth. These findings suggested the clinical utility of passive repetitive stretch to maintain muscle strength in unconscious or paralyzed patients and should be available for muscle training in sports and clinical rehabilitation.

Conclusion

Mechanical stimulation on muscle has very important role for muscle growth-hypertrophy and strengthening. Mechanical stimulation converted by mechano-sensor on the surface of muscle cell into chemical signaling and activate intracellular pathway to activate gene expression. Muscle strengthen exercise is thought to be based on these mechanism. Furthermore, passive stretching activates the same pathway on skeletal muscle. These findings are useful on clinical rehabilitation and sports training.

References

[1] Narici MV, Landoni L, Minetti AE: Assessment of human knee extensor muscles stress from in vivo physiological cross-sectional area and strength measurements. Eur *J Appl .Physiol.* 1992. 65: 438-444

[2] Vandenburgh H, Kaufman S: In vitro model for stretch-induced hypertrophy of skeletal muscle. *Science* 1979, 203 : 265-8

[3] Holly RG, Barnett JG, Ashmore CR, Taylor RG, Mole PA Stretch-induced growth in chicken wing muscles: a new model of stretch hypertrophy. *Am J Physiol* 1980,238: C62-C71

[4] Alway SE, Gonyea WJ, Davis ME. Muscle fiber formation and fiber hypertrophy during the onset of stretch-over-load. *Am J Physiol* 1990, 259: C92-C102.

[5] Laurent GJ, Sparrow MP, Millward DJ. Turnover of muscle protein in the fowl. Changes in rates protein synthesis and breakdown during hypertrophy of the anterior and posterior latissimus dorsi muscle. *Biochem J* 1978,176: 407-417

[6] Osbaldeston NJ, Lee DM, Cox VM, Hesketh JE, Morrison JF, Blair GE, Goldspink DF. The temporal and cellular expression of *c-fos* and c-*jun* in mechanically stimulated muscle. *Biochem J* 1995, 308: 465-471

[7] Carson JA, Booth FW. Myogenin mRNA is elevated during rapid, slow, and maintenance phases of stretch-induced hypertrophy in chicken slow-tonic muscle. *Pflugers-Arch* 1998, 435 : 850-858

[8] Gregory P, Gagnon J, Essig DA, Reid SK, Prior G, Zak R. Differential regulation of actin and myosin isozyme synthesis in functionally overloaded skeletal muscle. *Biochem J* 1990, 265: 525-532

[9] Ikeda S, Yoshida A, Matayoshi S, Tanaka N. Repetitive stretch induces c-*fos* and Myogenin mRNA within several hours in rat removed skeletal muscle. *Arch Phys Med Rehabil* 2003 84 419-23.

[10] Tamkun JW, DeSimone DW, Fonda D, Patel RS, Buck C, Horwitz AF, Hynes RO. Structure of integrin a glycoprotein involved in the transmembrane linkage between fibronectin and actin. *Cell* 1986, 46:271-282

[11] Clark EA, Brugge JS. Integrins and signal transduction pathways: The road taken. *Science* 1995, 268: 233-239

[12] Fluck M, Carson JA, Gordon SE, Ziemiecki A, Booth FW. Focal adhesion proteins FAK and paxillin increase in hypertrophied skeletal muscle. *Am J Physiol* 1999, 277: C152-C162.

[13] Hu Y, Bock G, Wick G, Xu Q. Activation of PDGF receptor alpha in vascular smooth muscle cells by mechanical stress. *FASEB J* 1998,12: 1135-1142

[14] Sadoshima J, Takahashi T, Jahn L, Izumo S. Roles of mechano-sensitive ion channels, cytoskeleton, and contractile activity in stretch-induced immediate-early gene expression and hypertrophy of cardiac myocytes. *Proc Natl Acad Sci* 1992, 89: 9905-9909

[15] Yamazaki T, Komuro I, Kudoh S, Zou Y, Nagai R, Aikawa R, Uozumi H, Yazaki Y. Role of channels and exchangers in mechanical stretch-induced cardiomyocyte hypertrophy. *Circ Res* 1998, 82: 430-437

[16] Perrone CE, Fenwick-Smith D, Vandenburgh HH. Collagen and stretch modulate autocrine secretion of insulin-like growth factor-I and insulin-like growth factor binding proteins from differentiated skeletal muscle cells. *J Biol Chem* 1995, 270: 2099-2106

[17] Mitchell P, Steenstrup T, Hannon K. Expression of fibroblast growth factor family during postnatal skeletal muscle hypertrophy. *J Appl Phys* 1999 86: 313-319

[18] Goldberg AL: Role of insulin in work-induced growth of skeletal muscle. *Endocrinology* 1968, 83: 1071-1073

[19] Turner JD, Rotwein P, Novakofski J, Bechtel PJ. Induction of mRNA for IGF-I and IGF-II during growth hormone-stimulated muscle hypertrophy. *Am J Physiol* 1988, 255: E513-E517

[20] Hameed M, Lange KH, Andersen JL, Schjerling P, Kjaer M, Harridge SD, Goldspink G. The effect of recombinant human growth hormone and resistance training on IGF-I mRNA expression in the muscles of elderly men. *J Physiol.* 2004, 555:231-40.

[21] Ikeda S, Yoshida A, Matayoshi S, Horinouchi K, Tanaka N. Induction of myogenin messenger ribonucleic acid in rat skeletal muscle after 1 hour of passive repetitive stretching. Arch Phys Med Rehabil. 2004, 85:166-7.

In: New Messenger RNA Research Communications
Editor: Lee B. Kwang, pp. 121-131

ISBN 978-1-60021-488-2

Chapter IX

New Research Communications on Messenger RNA Specific Inhibition of 801-Base Pair Deletion-Mutant *Epidermal Growth Factor Receptor* mRNA Expression Using Ribozymes

Marc-Eric Halatsch[7]
Department of Neurosurgery
Ruprecht Karls University
Heidelberg, Germany

ABSTRACT

The epidermal growth factor receptor (EGFR) is frequently amplified and mutated in human cancers and has been identified as a promising target for new anticancer strategies. Overexpression of the EGFR by tumor cells is associated with increased cellular proliferation, motility, adhesion, tumorigenicity and invasive capacity, induced angiogenesis as well as inhibition of apoptosis. Targeting strategies directed against wild-type or mutant EGFR include small molecule tyrosine kinase inhibitors, monoclonal antibodies, immunotoxin conjugates and various classes of antisense oligonucleotides. Among several mutations of the *EGFR*, a specific 801-base pair deletion mutation (termed variant III [vIII] or Δ*EGFR*) involving the extracellular receptor domain commonly occurs in various malignant neoplasms but not in normal tissues and thus represents an outstanding molecular target. The truncated receptor is constitutively (ligand-independently) activated and defective in downregulation due to low rates of receptor endocytosis. This chapter focuses on the promise and potential limitations of

[7]Address for correspondence: Marc-Eric Halatsch, M.D., Ph.D., Department of Neurosurgery, Ruprecht Karls University, Im Neuenheimer Feld 400, D-69120 Heidelberg, Germany. Phone: +49 6221 5639671; Fax: +49 6221 565534; E-mail: marc.halatsch@med.uni-heidelberg.de

specific ribozyme-mediated inhibition of Δ*EGFR* messenger (m)RNA expression as a gene therapeutic approach to ΔEGFR-expressing human malignancies.

Key Words: deletion-mutant epidermal growth factor receptor; ribozymes; RNA interference; retrovirus-mediated DNA transfer; tumorigenicity

INTRODUCTION

Within various cell types, the EGFR plays a pivotal role in transducing differentiation, developmental and mitogenic signals upon binding of its ligands EGF or transforming growth factor-α. Located on the short arm of chromosome 7 (p11-p13), the EGFR gene is 110 kb in size and has 26 exons. The gene encodes a 170 kDalton single-chain transmembrane glycoprotein spanning three functional domains, *i.e.*, an extracellular ligand-binding domain, a transmembrane domain and an intracellular tyrosine kinase domain. Rearrangements of the EGFR gene result in several mutants which have been detected in various human tumors. One specific rearrangement eliminates a DNA fragment containing exons 2-7 of the gene [35, 42] and results in an in-frame deletion of 801 base pairs (bp) (nucleotide [nt] positions 275-1075 [38]) of the coding sequence of the extracellular receptor domain (figure 1).

wt*EGFR* gene sequence
5'-...CTG GAG GAA AAG AAA **G***TT TGC CAA GGC ACG ... GTG AAG AAG TGT CCC C***GT** AAT TAT GTG GTG ACA GAT...-3'

Δ*EGFR* gene sequence
5'-...CTG GAG GAA AAG AAA GGT AAT TAT GTG GTG ACA GAT...-3'

Figure 1. Schematic representation of wt and ΔEGFR gene sequences. Italics, gene sequence subjected to deletion mutation; bold, newly generated codon; double underlining, target codon for ribozyme cleavage. Adapted and reproduced with permission from Halatsch et al., J Neurosurg 2000; 92: 297-305.

The deletion has been termed variant III (vIII) or Δ*EGFR* and brings together what were normally distant nucleotides in the intact gene, thus forming a unique target sequence. This specific type of structural alteration, being considered the most prevalent mutation involving the *EGFR* gene, is present in up to 58% of glioblastoma multiforme (GBM) [40], the most common malignant intrinsic brain tumor in adults for whom even aggressive multimodality treatment is almost always only palliative. The truncated EGFR is constitutively (ligand-independently) activated, defective in down-regulation due to low rates of receptor endocytosis [20] and has been shown to confer enhanced tumorigenicity [29] by increasing proliferation and reducing apoptosis *in vivo* [27]. This is consistent with the observation that its presence is prognostic of a shorter interval to clinical relapse and poorer survival in patients with GBM [34]. Quantitative and qualitative aspects thus render this mutated receptor a potentially important therapeutic target [12-16]. The ΔEGFR has several features in common with the principal oncogene of avian erythroblastosis virus, v-*erb*B [9], which arose from retroviral insertion that reproducibly deleted most of the ectodomain of the avian EGFR [28]. In contrast, the gene deletion characteristic for ΔEGFR occurs naturally during malignant progression [6, 34, 41]. Evidence suggests that ΔEGFR couples to the mitogen-activated protein-kinase signaling pathway which is also utilized by wild-type (wt)EGFR [32].

As a result of the deletion mutation, the fusion junction of the gene is created directly upstream of a GTA triplet which is subsequently transcribed into a ribozyme target codon (GUA).

RIBOZYMES

Ribozymes are catalytically active subsets of small RNA molecules which possess the dual properties of RNA sequence-specific recognition, analogous to conventional antisense molecules, and site-specific cleavage of RNA substrates [19], thereby intercepting gene expression and forestalling subsequent translation. While cleavage is mediated by the ribozyme's catalytic core, substrate recognition occurs through complementary flanking sequences. The fusion junction of *ΔEGFR* mRNA is unique in that a search of GenBank did not identify this sequence in any known human gene, and the flexibility of a ribozyme's flanking sequences thus allows specific targeting of this region.

Among several naturally occuring ribozymes that have been identified [23], the hammerhead and hairpin types are the best characterized. Hairpin ribozymes are larger than hammerhead ribozymes and possess about the same catalytic efficiency at lower magnesium concentration requirements [2], which may be advantageous for hairpin ribozyme activity *in vivo*. Three different hairpin ribozyme motifs (derived from the negative strands of satellite RNAs from tobacco ringspot virus [sTRSV], chicory yellow mottle virus [sCYMV1; figure 2] and arabis mosaic virus [sArMV]), have so far been described [5].

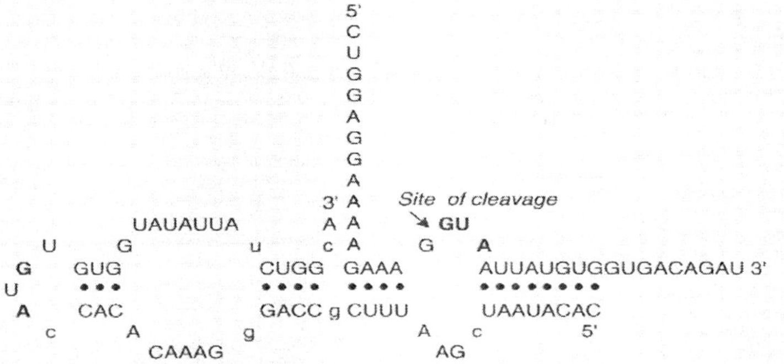

Figure 2. Schematic representation of the molecular interaction between the sCYMV1-derived hairpin ribozyme and ΔEGFR mRNA. Lower-case letters, base exchanges; bold letters, additional bases (compared with the "standard" sTRSV-derived ribozyme). Adapted and reproduced with permission from Halatsch et al., J Neurosurg 2000; 92: 297-305.

SPECIFIC INHIBITION OF Δ*EGFR* MRNA EXPRESSION BY RIBOZYME-TARGETING

Two experimental studies [11, 43] have addressed the question of whether a ribozyme specifically targeting Δ*EGFR* could achieve reversal of the ΔEGFR-associated malignant phenotype in a molecularly defined subgroup of GBM (figure 3).

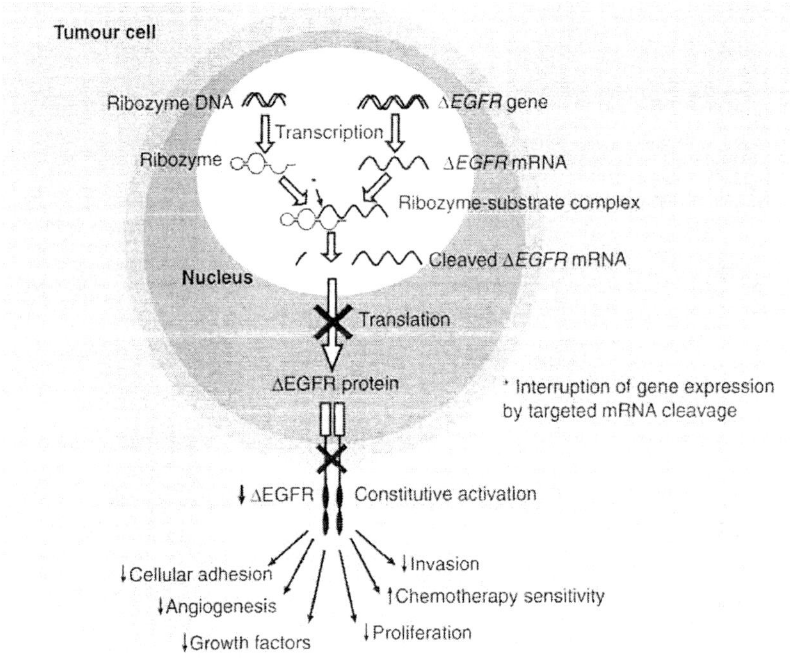

Figure 3. Schematic representation of anticancer effects resulting from ribozyme-mediated inhibition of ΔEGFR expression by targeted mRNA cleavage. Reproduced with permission from Halatsch, Signal 2001; 2: 12-16.

In both studies, ribozymes were designed to cleave the GUA sequence located directly downstream of the fusion junction in Δ*EGFR* mRNA. The target codon is located within the sequence coding for the receptor's extracellular domain; considering the ligand-independent activation to be a result of the deletion mutation, this locus may be regarded as essential for autonomous EGFR function. In addition, the mRNA 5'-fragment carrying the translation initiation codon comprises only exon 1; thus it appears reasonable to assume that the cleavage products are non-functional even if translated into polypeptides. In a fine study, Yamazaki *et al.* have demonstrated specific hammerhead ribozyme-mediated cleavage of a synthetic Δ*EGFR* RNA substrate in a cell-free system. Their ribozyme exhibited efficient catalytic activity at Mg^{2+} concentrations as low as 1 mM. Although the cleavage reaction has an absolute requirement for Mg^{2+} and the intracellular Mg^{2+} concentration appears suboptimal for ribozyme activity, other divalent catalysts, *e.g.*, Ca^{2+}, also promote formation of a catalytically proficient ribozyme-substrate complex *in vivo* [2]. Subsequently, this group constructed a hammerhead ribozyme expression plasmid which was introduced into ERM5-1 cells (*i.e.*, murine NIH/3T3 cells transfected to express Δ*EGFR* cDNA) *in vitro* using calcium phosphate coprecipitation. The transfected ribozyme suppressed expression of Δ*EGFR* mRNA and reduced the growth of tumors formed by transduced ERM5-1 cells in nude mice. The authors also showed that the mitotic activity of ribozyme-expressing ERM5-1 cells as measured by bromodeoxyuridine incorporation was decreased *in vivo*.

In another study [11], three different hairpin ribozymes were designed to specifically target Δ*EGFR* mRNA, and the sCYMV1 motif was identified in a cell-free system as a promising candidate to be further evaluated as a potential agent against Δ*EGFR*, given the cleavage efficacy at physiological pH, temperature, and Mg^{2+} concentration as well as the non-interference with wt*EGFR* which assumes an essential role in normal cells. Detectable cleavage occurred at only 0.1 mM $MgCl_2$ which equals approximately 20% of the intracellular Mg^{2+} concentration. That particular hairpin ribozyme motif was then cloned into the amphotropic retroviral vector N2A+$tRNA_i^{met}$ and transferred into human U-87MG.ΔEGFR glioblastoma cells which overexpress exogenous ΔEGFR.

N2A+$tRNA_i^{met}$ was chosen for this application for several reasons. Among the particular advantages of this vector is the extreme transcription efficacy of the inserted DNA sequence which is due to the approximately 100-fold abundance of pol III-transcribed tRNA molecules as compared to polyadenylated RNA on a molar basis, and the placement of the foreign gene outside of the retroviral transcriptional unit, thereby eliminating or at least reducing the negative effects which may be exerted by this viral unit on the activity of the internal promoter [18, 36]. Moreover, the applicability of pol III transcription termination sequences assures that non-ribozyme sequences which could interfere with the ribozyme's secondary and tertiary structure are not added to its 3'-end, and finally, the chimeric tRNA-ribozyme design [37] confers on the transcript relative resistance against intracellular degradation [3, 44]. The latter issue is of particular relevance since successive rounds of substrate binding, cleavage, and product release typically enable a single ribozyme molecule to process more than one substrate molecule [23].

In this study, both transduced (including control infections) and non-transduced U-87MG.ΔEGFR cells were resistant against neomycin, and, analogous to a potential clinical situation, drug selection was not employed to eradicate non-transduced cells. Consequently, the data obtained on proliferation, clonogenicity, and *EGFR* mRNA expression did not reflect properties of a purely transduced cell population, but of presumably heterogenous populations consisting of transduced as well as a small fraction of non-transduced cells. The experimental conditions under which these mixed populations were generated included a virus:cell ratio of 40:1. Despite the heterogeneity, provirus integration and expression levels were similar among ribozyme-, disabled ribozyme- and mock-transduced cells; therefore, the observed differences in biological activity were unlikely secondary to diffences in provirus integration and expression. As in the study by Yamazaki *et al.* [43], the catalytically disabled ribozyme's lack of biological activity suggests that the inhibition of proliferation and anchorage-independent growth exerted by the original ribozyme on U-87MG.ΔEGFR cells was due to ribozyme-specific catalytic cleavage rather than due to an antisense effect.

Transduction of U-87MG.ΔEGFR with anti-Δ*EGFR*-sCYMV1 under relative serum starvation conditions amounted to 69% inhibition of ΔEGFR-mediated proliferation advantage. Yet, clonogenicity in soft agar or anchorage-independent growth (under identical relative serum starvation conditions) which serves as the best *in vitro* correlate to tumorigenicity was profoundly (>95%) suppressed by transfer of anti-Δ*EGFR*-sCYMV1. Taken together, it appears that in these test systems ΔEGFR affects the interaction of GBM cells with their environment rather than cell proliferation directly and that these ΔEGFR-mediated effects may be specifically and efficiently intercepted by a targeted ribozyme, the growth-suppressing action of which is intensified in a three-dimensional environment. Noteworthy, overexpression of the truncated EGFR in U-87MG.ΔEGFR is comparable to levels seen in primary brain tumors [27].

Retroviral vectors only integrate and subsequently express their genes in proliferating tissues (*e.g.*, tumors), and in terms of transduction specificity this feature is considered advantageous in the brain where the tumor is the predominant mitotic cell type, thus maximizing selective transduction of tumor with minimal or absent transduction of the non-dividing normal brain cells [4]. Moreover, the vector-mediated selectivity is complemented by the specificity of the molecular targeting mechanism which was corroborated in both cell-free and cellular systems.

The findings of both studies indicate that anti-Δ*EGFR* ribozymes are capable of specifically inhibiting the expression of Δ*EGFR* and reversing the Δ*EGFR*-associated malignant phenotype of GBM cells. Thus, this strategy may constitute a promising gene therapeutic approach for a molecularly defined subgroup of GBM and other ΔEGFR-expressing human cancers (*e.g.*, subgroups of carcinomas of the lung [10], breast [26, 39], ovary [26] and prostate [31]) as well. Indeed, by means of a targeted hammerhead ribozyme, Luo *et al.* [24] have achieved suppression of proliferation and tumorigenicity of breast cancer cells expressing low levels of endogenous Δ*EGFR*. Although their method of ribozyme delivery, relying on calcium phosphate coprecipitation and subsequent selection by neomycin, is far from any realistic clinical application, this study provides proof-of-principle for ribozyme-mediated targeting of Δ*EGFR* mRNA in yet another neoplasm besides glioblastoma.

Conclusion

Ribozyme-mediated targeting of Δ*EGFR* expression as an anticancer strategy appears attractive due to the unsurpassed specificity of the approach. At the same time, the pharmacokinetic limitations (*i.e.*, low transduction efficiency) which became evident during the *herpes simplex virus thymidine kinase* gene/ganciclovir (HSV TK/GCV) trial on GBM apply to ribozyme-mediated targeting of Δ*EGFR* expression as well. On the other hand, the target specificity of the effector molecule may allow for extended and new applications of the anti-Δ*EGFR* ribozyme approach in settings where systemic toxicity might otherwise be a concern (*e.g.*, delivery by extremely high-titered or replication-competent vectors, if those vectors do not carry a systemic risk within themselves; local injection even in situations prone to systemic spread, such as after inadvertent opening of the lateral ventricles during radical (re-)resection of GBM). Compared to the HSV TK/GCV system, the anti-Δ*EGFR* ribozyme approach appears to elicit a cytostatic rather than a cytocidal effect but may increase susceptibility of targeted cancer cells towards conventional and new therapeutic modalities and agents.

Alternatively, despite the fact that multiple genetic alterations underly the initiation and progression of cancer [8], this approach may by itself possess the potential to trigger apoptosis by disrupting the constitutively active signaling of ΔEGFR. It is important to keep in mind that existing research in this field was conducted with (initially) cultured target cells upon which ΔEGFR expression had to be forced by transfection or retrovirus-mediated gene transfer since these cells usually do not express the mutated receptor under standard cell culture conditions. Thus, in contrast to naturally occurring ΔEGFR-expressing cancer cells *in vivo*, these engineered target cells are unlikely to be highly dependent upon constitutive ΔEGFR signaling, a fact that could nourish hope that ribozyme-mediated inhibition of ΔEGFR expression might be even more effective in a "natural" setting (figure 4).

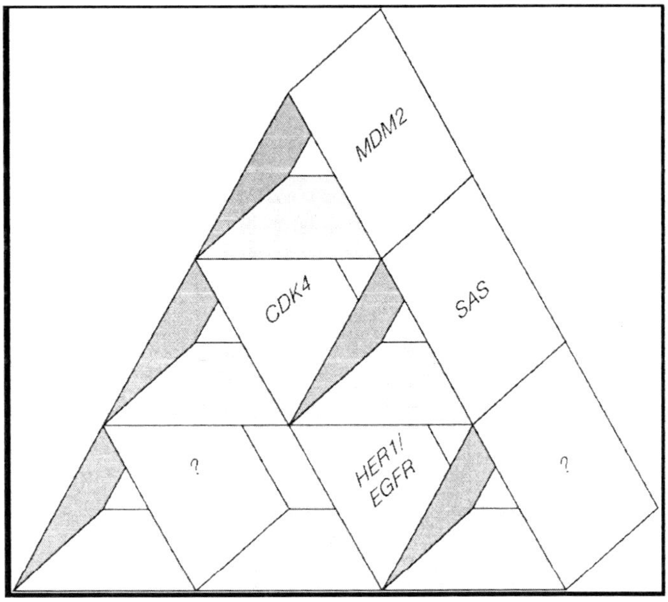

Figure 4. Antiapoptotic balance in glioblastoma is a house of (gene) cards. Eliminating the contribution of EGFR aberrant signalling by therapeutic targeting results in critical destabilization. Abbreviations: CDK4, cyclin-dependent kinase 4, HER1/EGFR, epidermal growth factor receptor; MDM2, murine double minute 2; SAS, sarcoma-amplified sequence. Reproduced with permission from Halatsch et al., Cancer Treat Rev 2006; 32: 74-89.

Recently, a mechanism called RNA interference (RNAi) has emerged as a powerful tool for gene silencing [7]. In RNAi, exogenous double-stranded RNAs are cleaved into fragments of 21-23 nt with 2-nt 3' overhangs (short interfering [si]RNAs), and these are subsequently incorporated into the RNA-induced silencing complex (RISC). The siRNA-RISC complex recognizes and cleaves the target RNA in a sequence-specific manner [17, 30]. The most important difference between ribozymes and siRNAs is that the latter involve recruitment of endogenous proteins while the actions of ribozymes are independent of these intracellular factors [33]. Thus, because siRNAs engage a vast array of intracellular enzymes, they are more effective than ribozymes in the knockdown of target mRNAs [1]. However, a study on combination of siRNAs with ribozymes revealed consistent improvement of siRNA inhibitory activity by a ribozyme in the range of 25% independent of the level of inhibition already achieved by different siRNAs [21]. Off-target effects, including induction of interferon responses, are more frequently observed with the use of siRNA, making ribozymes seem superior in terms of specificity of action [22]. Moreover, RNAi targets RNA molecules primarily in the cytoplasm of mammalian cells whereas ribozymes can be designed to target introns and nuclear-localized RNA. Thus, in general, ribozymes may be more useful when it is necessary to selectively downregulate a sequence derived from from a gene family of highly homologous sequences in which only the introns have grossly different sequences. In addition, ribozymes can be used under circumstances where it is highly desirable to cleave mRNA before it reaches the cytoplasm, *e.g.*, to inhibit human immunodeficiency virus replication [25].

Taken together, while it is anticipated that the use of siRNA will continue to expand, RNAi appears unlikely to supplant the use of ribozymes and related approaches for many applications. Rather, the choice of antisense methodology will depend on the specific demands of the intended use. Regardless of the effector methodology under consideration, patients, clinicians and researchers are impatiently awaiting the next generation of highly efficient delivery systems.

REFERENCES

[1] Beale G., Hollins A. J., Benboubetra M., Sohail M., Fox S. P., Benter I. and Akhtar S. (2003). Gene-silencing nucleic acids designed by scanning arrays: anti-EGFR activity of siRNA, ribozyme and DNA enzymes targeting a single hybridization-accessible region using the same delivery system. *J Drug Target, 11,* 449-456.

[2] Chowrira B. M., Berzal-Herranz A. and Burke J. M. (1993). Ionic requirements for RNA binding, cleavage, and ligation by the hairpin ribozyme. *Biochemistry, 32,* 1088-1095.

[3] Cotten M. and Birnstiel M. L. (1989). Ribozyme-mediated destruction of RNA *in vivo*. *EMBO J, 8,* 3861-3866.

[4] Culver K. W., Van Gilder J., Link C. J., Carlstrom T., Buroker T., Yuh W., Koch K., Schabold K., Doornbas S., Wetjen B., *et al.* (1994). Gene therapy for the treatment of malignant brain tumors with *in vivo* tumor transduction with the *herpes simplex thymidine kinase* gene/ganciclovir system. *Human Gene Ther, 5,* 343-379.

[5] DeYoung M., Siwkowski A. M., Lian Y. and Hampel A. (1995). Catalytic properties of hairpin ribozymes derived from Chicory yellow mottle virus and arabis mosaic virus satellite RNAs. *Biochemistry, 34,* 15785-15791.

[6] Ekstrand A. J., Sugawa N., James C. D. and Collins V. P. (1992). Amplified and rearranged *epidermal growth factor receptor* genes in human glioblastomas reveal deletions of sequences encoding portions of the N- and/or C-terminal tails. *Proc Natl Acad Sci U S A, 89,* 4309-4313.

[7] Fire A., Xu S., Montgomery M. K., Kostas S. A., Driver S. E. and Mello C. C. (1998). Potent and specific genetic interference by double-stranded RNA in *Caenorhabditis elegans*. *Nature, 391*, 806-811.

[8] Fischer U., Meltzer P. and Meese E. (1996). Twelve amplified and expressed genes localized in a single domain in glioma. *Hum Genet, 98,* 625-628.

[9] Frykberg L., Palmieri S., Beug H., Graf T., Hayman M. J. and Vennstrom B. (1983). Transforming capacities of avian erythroblastosis virus mutants deleted in the *erb*A or *erb*B oncogenes. *Cell, 32,* 227-238.

[10] Garcia de Palazzo I. E., Adams G. P., Sundareshan P., Wong A. J., Testa J. R., Bigner D. D. and Weiner L. M. (1993). Expression of mutated epidermal growth factor receptor by non-small cell lung carcinomas. *Cancer Res, 53,* 3217-3220.

[11] Halatsch M.-E., Schmidt U., Bötefür I. C., Holland J. F. and Ohnuma T. (2000). Marked inhibition of glioblastoma target cell tumorigenicity *in vitro* by retrovirus-mediated transfer of a hairpin ribozyme against deletion-mutant *epidermal growth factor receptor* messenger RNA. *J Neurosurg, 92,* 297-305.

[12] Halatsch M.-E. (2001). Selective ribozyme-mediated inhibition of 801-bp deletion-mutant *EGFR* mRNA expression in human cancers – can a war be won by killing one messenger? *Signal, 2,* 12-16.

[13] Halatsch M.-E., Schmidt U., Bötefür I. C., Holland J. F. and Ohnuma T. (2001). Overexpression of deletion-mutant epidermal growth factor receptor is associated with altered genotoxic stress-provoked *p*53 mRNA induction in a human glioblastoma cell line. *Anticancer Res, 21,* 189-196.

[14] Halatsch M.-E., Gehrke E. E., A.-Borhani F., Efferth T., Gebhart E., Domhof S., Schmidt U. and Buchfelder M. (2003). EGFR but not PDGFR-β expression correlates to the antiproliferative effect of growth factor withdrawal in glioblastoma multiforme cell lines. *Anticancer Res, 23,* 2315-2320.

[15] Halatsch M.-E., Gehrke E. E., Vougioukas V. I., Bötefür I. C., A.-Borhani F., Efferth T., Gebhart E., Domhof S., Schmidt U. and Buchfelder M. (2004). Inverse correlation of *epidermal growth factor receptor* (*EGFR*) mRNA induction and suppression of anchorage-independent growth by OSI-774, an EGFR tyrosine kinase inhibitor, in glioblastoma multiforme cell lines. *J Neurosurg, 100*, 523-533.

[16] Halatsch M.-E., Schmidt U., Behnke-Mursch J., Unterberg A. and Wirtz C. R. (2006). Epidermal growth factor receptor inhibition for the treatment of glioblastoma multiforme and other malignant brain tumours. *Cancer Treat Rev, 32,* 74-89.

[17] Hammond S. M., Boettcher S., Caudy A. A., Kobayashi R. and Hannon G. J. (2001). Argonaute2, a link between genetic and biochemical analyses of RNAi. *Science, 293,* 1146-1150.

[18] Hantzopoulos P. A., Sullenger B. A., Ungers G. and Gilboa E. (1989). Improved gene expression upon transfer of the *adenosine deaminase* minigene outside the transcriptional unit of a retroviral vector. *Proc Natl Acad Sci U S A, 86,* 3519-3523.

[19] Haseloff J. and Gerlach W. L. (1988). Simple RNA enzymes with new and highly specific endoribonuclease activities. *Nature, 334,* 585-591.

[20] Huang H. S., Nagane M., Klingbeil C. K., Lin H., Nishikawa R., Ji X. D., Huang C. M., Gill G. N., Wiley H. S. and Cavenee W. K. (1997). The enhanced tumorigenic activity

of a mutant epidermal growth factor receptor common in human cancers is mediated by threshold levels of constitutive tyrosine phosphorylation and unattenuated signaling. *J Biol Chem, 272,* 2927-2935.

[21] Jarczak D., Korf M., Beger C., Manns P. and Krüger M. (2005). Hairpin ribozymes in combination with siRNAs against highly conserved hepatitis C virus sequence inhibit RNA replication and protein translation from hepatitis C virus subgenomic replicons. *FEBS J, 272,* 5910-5922.

[22] Lewin A. S., Drenser K. A., Hauswirth W. W., Nishikawa S., Yasumura D., Flannery J. G. and LaVail M. M. (1998). Ribozyme rescue of photoreceptor cells in a transgenic rat model of autosomal dominant retinitis pigmentosa. *Nature Med, 4,* 967-971.

[23] Long D. M. and Uhlenbeck O. C. (1993). Self-cleaving catalytic RNA. *FASEB J, 7,* 25-30.

[24] Luo X., Gong X. and Tang C. K. (2003). Suppression of EGFRvIII-mediated proliferation and tumorigenesis of breast cancer cells by ribozyme. *Int J Cancer, 104,* 716-721.

[25] Michienzi A., Cagnon L., Bahner I. and Rossi J. J. (2000). Ribozyme-mediated inhibition of HIV-1 suggests nucleolar trafficking of HIV-1 RNA. *Proc Natl Acad Sci U S A, 97,* 8955-8960.

[26] Moscatello D. K., Holgado-Madruga M., Godwin A. K., Ramirez G., Gunn G., Zoltick P. W., Biegel J. A., Hayes R. L. and Wong A. J. (1995). Frequent expression of a mutant epidermal growth factor receptor in multiple human tumors. *Cancer Res, 55,* 5536-5539.

[27] Nagane M., Coufal F., Lin H., Bögler O., Cavenee W. K. and Huang H. J. (1996). A common mutant epidermal growth factor receptor confers enhanced tumorigenicity on human glioblastoma cells by increasing proliferation and reducing apoptosis. *Cancer Res, 56,* 5079-5086.

[28] Nilsen T. W., Maroney P. A., Goodwin R. G., Rottman F. M., Crittenden L. B., Raines M. A. and Kung H. J. (1985). c-*erb*B activation in ALV-induced erythroblastosis: novel RNA processing and promoter insertion result in expression of an amino-truncated EGF receptor. *Cell, 41,* 719-726.

[29] Nishikawa R., Ji X.-D., Harmon R. C., Lazar C. S., Gill G. N., Cavenee W. K. and Huang H. J. (1994). A mutant epidermal growth factor receptor common in human glioma confers enhanced tumorigenicity. *Proc Natl Acad Sci U S A, 91,* 7727-7731.

[30] Nykanen A., Haley B. and Zamore P. D. (2001). ATP requirements and small interfering RNA structure in the RNA interference pathway. *Cell, 107,* 309-321.

[31] Olapade-Olaopa E. O., Moscatello D. K., MacKay E. H., Horsburgh T., Sandhu D. P., Terry T. R., Wong A. J. and Habib F. K. (2000). Evidence for the differential expression of a variant EGF receptor protein in human prostate cancer. *Br J Cancer, 82,* 186-194.

[32] Pawson T. (1995). Protein modules and signalling networks. *Nature, 373,* 573-580.

[33] Scherer, L. J. and Rossi J. J. (2003). Approaches for the sequence-specific knockdown of mRNA. *Nature Biotechnol, 21,* 1457-1465.

[34] Schlegel J., Merdes A., Stumm G., Albert F. K., Forsting M., Hynes N. and Kiessling M. (1994). Amplification of the *epidermal growth factor receptor* gene correlates with different growth behaviour in human glioblastoma. *Int J Cancer, 56,* 72-77.

[35] Sugawa N., Ekstrand A. J., James C. D. and Collins V. P. (1990). Identical splicing of aberrant *epidermal growth factor receptor* transcripts from amplified rearranged genes in human glioblastomas. *Proc Natl Acad Sci U S A, 87,* 8602-8606.

[36] Sullenger B. A., Lee T. C., Smith C. A., Ungers G. E. and Gilboa E. (1990). Expression of chimeric tRNA-driven antisense transcripts renders NIH 3T3 cells highly resistant to Moloney murine leukemia virus replication. *Mol Cell Biol, 10,* 6512-6523.

[37] Thompson J. D., Ayers D. F., Malstrom T. A., McKenzie T. L., Ganousis L., Chowrira B. M., Couture L. and Stinchcomb D. T. (1995). Improved accumulation and activity of ribozymes expressed from a tRNA-based RNA polymerase III promoter. *Nucl Acid Res, 23,* 2259-2268.

[38] Ullrich A., Coussens L., Hayflick J. S., Dull T. J., Gray A., Tam A. W., Lee J., Yarden Y., Liberman T. A., Schlessinger J., *et al.* (1984). Human *epidermal growth factor receptor* cDNA sequence and aberrant expression of the amplified gene in A431 epidermoid carcinoma cells. *Nature, 309,* 418-425.

[39] Wikstrand C. J., Hale L. P., Batra S. K., Hill M. L., Humphrey P. A., Kurpad S. N., McLendon R. E., Moscatello D., Pegram C. N., Reist C. J., *et al.* (1995). Monoclonal antibodies against EGFRvIII are tumor-specific and react with breast and lung carcinomas and malignant gliomas. *Cancer Res, 55,* 3140-3148.

[40] Wikstrand C. J., McLendon R. E., Friedman A. H. and Bigner D. D. (1997). Cell surface localization and density of the tumor-associated variant of the epidermal growth factor receptor, EGFRvIII. *Cancer Res, 57,* 4130-4140.

[41] Wong A. J., Ruppert J. M., Bigner S. H., Grzeschik C. H., Humphrey P. A., Bigner D. S. and Vogelstein B. (1992). Structural alterations of the *epidermal growth factor receptor* gene in human gliomas. *Proc Natl Acad Sci U S A, 89,* 2965-2969.

[42] Yamazaki H., Ohba Y., Tamaoki N. and Shibuya M. (1990). A deletion mutation within the ligand-binding domain is responsible for activation of the *epidermal growth factor receptor* gene in human brain tumors. *Jpn J Cancer Res, 81,* 773-779.

[43] Yamazaki H., Kijima H., Ohnishi Y., Abe Y., Oshika Y., Tsuchida T., Tokunaga T., Tsugu A., Ueyama Y., Tamaoki N. and Nakamura M. (1998). Inhibition of tumor growth by ribozyme-mediated suppression of aberrant *epidermal growth factor receptor* gene expression. *J Natl Cancer Inst, 90,* 581-587.

[44] Yuyama N., Ohkawa J., Inokuchi Y., Shirai M., Sato A., Nishikawa S. and Taira K. (1992). Construction of a tRNA-embedded ribozyme trimming plasmid. *Biochem Biophys Res Comm, 186,* 1271-1279.

In: New Messenger RNA Research Communications
Editor: Lee B. Kwang, pp. 133-166
ISBN 978-1-60021-488-2

Chapter X

A Bioinformatical Approach to the Analysis of Viral and Cellular Internal Ribosome Entry Sites

Martin Mokrejš[1,*], Václav Vopálenský[1], Tomáš Mašek[1] and Martin Pospíšek[1,2,*]

[1]Department of Genetics and Microbiology, Charles University, Faculty of Science, Viničná 5, 128 43 Prague, Czech Republic

[2]Centre of Biomedical Informatics, Institute of Computer Science AS CR, Pod Vodárenskou věží 2, 182 07 Prague 8, Czech Republic

ABSTRACT

The internal ribosome entry site (IRES) is a part of the mRNA sequence which is able to attract the eukaryotic ribosomal initiation complex and to directly promote the initiation of protein synthesis independently of the presence of 5'-terminal 7mG cap. RNA structures bearing the IRES activity were first discovered in certain eukaryotic viruses where they very often play a pivotal role in viral strategies, allowing the viral invader to overcome the overall decrease of the host protein synthesis caused either by viral proteins or by the cellular antiviral defense system. Although the IRES segments and thus the cap-independent translation initiation were first described in viruses, extensive evidence has appeared in the past few years that a similar principle of the translation initiation is utilized also by some cellular mRNAs. Demonstration of IRES activity of a particular RNA region is not a simple task. A proper design of the experiment and a careful selection of the controls – excluding artificial signals generated by leaky scanning, ribosome hopping and undesirable cryptic transcription, splicing or physical breakage at the hot-spots – are very important. A number of false positives described in the literature as well as difficulties in designing appropriate controls have become the major stimuli for creating IRESite – the publicly available manually

[*] Corresponding authors: M. Mokrejš *mmokrejs@iresite.org*, M. Pospíšek *martin@natur.cuni.cz*

annotated database of experimentally verified IRES structures. This chapter presents the current status of the IRESite database (http://www.iresite.org), the complete list of known viral and cellular IRESs as well as novel results obtained from the comparative analyses of IRES segments accumulated to date. The article also presents a brief description and comparison of other available databases containing IRES and 5' untranslated region (5'-UTR) related information.

ABBREVIATIONS

7mG – 7-methylguanosine;
IRES – internal ribosome entry site;
ITAF – IRES trans-acting factor;
TE – translational enhancer; UTR – untranslated region

INTERNAL RIBOSOME ENTRY SITE (IRES) SEGMENTS FACILITATE INITIATION OF TRANSLATION IN EUKARYOTES

Eukaryotic protein synthesis is a highly regulated cellular process which consumes about 5% of total human caloric uptake (Mathews *et al.*, 2000). Kinetic studies of translation revealed that the rate-limiting step of protein synthesis is the initiation, during which the necessary translation factors have to build physical connection between the 5'-end of mRNA molecule and the 40S ribosome subunit to assemble a functional initiation complex. The whole process consists of a complicated series of interactions among the proteins, RNAs and low molecular weight molecules having a specific order and timing. Eukaryotic cells utilize a 7mG cap moiety to protect the 5'-end of the mRNA molecules against nuclease attack as well as to facilitate their translation by attachment of cap-binding translation initiation factors (Hershey and Merrick, 2000).

The mechanism of translation initiation of certain viral RNAs was discovered 20 years ago in the labs of E. Wimmer and N. Sonenberg (Dorner *et al.*, 1984; Jang *et al.*, 1988; Pelletier and Sonenberg, 1988). It appeared that eukaryotic ribosomes can – besides the canonical cap-dependent pathway – enter the mRNA transcript also independently of the 5'-cap structure in a region called *I*nternal *R*ibosome *E*ntry *S*ite (IRES). Some of the eukaryotic viruses use this uncommon mechanism of translation initiation as part of their life strategy allowing them to completely usurp the cellular translational machinery for preferential synthesis of viral proteins. However, the growing list of experimental data accumulated during the past decade provided evidence that the cap-independent translation initiation is not limited only to the particular viruses but can also be utilized by certain cellular mRNAs at least in cases when the cap-dependent translation initiation is compromised, as happens often during stress, hypoxia, viral infection, *etc.*

When thinking about the IRESs it must be stated first that the internal ribosome entry site is defined by functional assays on principle. Attempts to find some common sequence or structural criteria for the IRES have failed so far. Viral IRESs are usually much better characterized than cellular IRESs and some of them – including the prominent hepatitis C virus IRES – have been studied so intensively that the results contributed significantly to the current view of translation initiation. Viral IRESs are often highly structured and even if stronger similarities can be found only in closely related viruses, they can be classified by the operational criteria (Jackson, 2000). In the field of cellular IRESs the situation is currently less transparent. It is now becoming accepted that a remarkable part of the currently reported IRESs encoded by nuclear genes has been published without proper experimental controls. These findings gave rise to doubts about the existence of many reported cellular IRESs and prompted scientists to discuss extensively an introduction of new strict evaluation criteria for IRES determination and classification (Jackson, 2000; Kozak, 2005).

Several reviews and articles shedding light on IRESs from different perspectives have recently been published (Jackson, 2000; Kean, 2003; Merrick, 2004; Stoneley and Willis, 2004; Holcik and Sonenberg, 2005; Komar and Hatzoglou, 2005; Spriggs *et al.*, 2005).

We present herein an up-to-date compilation of viral, cellular and artificially constructed IRESs. To date, 56 viruses (Table 1 and 2) as well as 82 cellular genes (Table 3) have been reported to contain IRES in their 5'-untranslated regions. Several attempts to select IRES segments from the pool of random sequences led only to short sequences showing a high degree of complementarity to rRNA (Table 4). Many of the IRESs listed in Tables 1 to 4 have already been included in the IRESite database and became the source for the analyses presented in this study (Table 5).

As seen from Table 1, since the time of the first discovery of viral cap-independent mode of translation initiation the number of viruses known to utilize internal ribosome entry site has grown substantially. It is obvious that the use of any kind of internal ribosome entry site should be especially advantageous for single stranded +RNA viruses lacking the cap structure. Indeed the vast majority of the animal and human viruses containing IRES are RNA viruses or retroviruses including retrotransposons, and most of them belong only to the three families of +ssRNA viruses (*Picornaviridae*, *Dicistroviridae* and *Flaviviridae*). A mere six DNA viruses have so far been reported to contain IRES, four of them belonging to the *Herpesviridae* family (Table 1).

The translation initiation strategy of many plant viruses relies on translational enhancers (TE) at their 3'-UTR or 5'-UTR, which are often based on complementarity to 18S rRNA, or on 5' to 3' ends crosstalk facilitated by the kissing stem loop structures (Table 2). Translational control in plant +RNA viruses has been reviewed recently (Dreher and Miller, 2006; Kneller *et al.*, 2006).

Table 3 presents the list of known IRESs found in 5'-UTRs of cellular mRNAs. The table does not contain any IRESs or translational enhancers located in 3'-UTR. It should be noted that only few IRESs listed in Table 3 have been studied more extensively and some of them have been claimed afterwards not to be IRES at all. The latter are notably TIF4631 (Zhou *et al.*, 2001; Altmann *et al.*, 2004; Mauro *et al.*, 2004; Verge *et al.*, 2004), HAP4, TFIID, YAP1 (Hecht *et al.*, 2002; Seino *et al.*, 2005), eIF4G (Han and Zhang, 2002), c-SIS/PDGF2 (Han *et al.*, 2003); $p27^{kip}$ (Liu *et al.*, 2005) and Pim-1 (Wang *et al.*, 2005) which have been reported to contain a promoter in their cDNA copy. The Gtx, RBM3 and RPS18C genes and the spliced leader (SL) RNA from *Leishmania* contain a region complementary to ribosomal RNA of the small ribosomal subunit. All reported IRES segments selected from the pool of random sequences *in vitro* also fall into the same group of IRESs (Table 4).

Table 1. List of animal and human viruses containing IRES (51 in total).

Virus abbreviation	Full name of the virus	Family	References
ALPV	Aphid lethal paralysis virus	*Dicistroviridae*	(van Munster *et al.*, 2002)
BEV	Bovine enterovirus	*Picornaviridae*	(Zell *et al.*, 1999)
BVDV	Bovine viral diarrhea virus	*Flaviviridae*	(Poole *et al.*, 1995)
CSFV	Classical swine fewer virus	*Flaviviridae*	(Rijnbrand *et al.*, 1997)
CVB3	Coxsackie virus B3	*Picornaviridae*	(Yang *et al.*, 1997)
CrPV	Cricket paralysis virus	*Dicistroviridae*	(Wilson *et al.*, 2000)
DCV	Drosophila C virus	*Dicistroviridae*	(Cherry *et al.*, 2005)
E6	Echovirus 6	*Picornaviridae*	(Beaulieux *et al.*, 2005)
E11	Echovirus 11	*Picornaviridae*	(Gharbi *et al.*, 2006)
E25	Echovirus 25	*Picornaviridae*	(Bailly *et al.*, 1996)
EAV	Equine arteritis virus	*Arteriviridae*	(Tijms *et al.*, 2001)
EBV	Epstein-Barr virus	*Herpesviridae*	(Isaksson *et al.*, 2003)
EMCV	Encephalomyocarditis virus	*Picornaviridae*	(Jang *et al.*, 1988)
ERAV	Equine rhinitis A virus	*Picornaviridae*	(Hinton *et al.*, 2000)
ERBV	Equine rhinitis B virus	*Picornaviridae*	(Hinton and Crabb, 2001)
EV71	Enterovirus 71	*Picornaviridae*	(Lee *et al.*, 2005)
FMDV	Foot-and-mouth disease virus	*Picornaviridae*	(Belsham and Brangwyn, 1990; Kuhn *et al.*, 1990)
F-MLV	Friend murine leukemia virus (mouse retrotransposon)	*Retroviridae*	(Berlioz and Darlix, 1995)
GBV-A	Hepatitis G virus A	*Flaviviridae*	(Simons *et al.*, 1996)
GBV-B	Hepatitis G virus B	*Flaviviridae*	(Grace *et al.*, 1999)
GBV-C	Hepatitis G virus C	*Flaviviridae*	(Simons *et al.*, 1996)
GLV	Giardia lamblia virus	*Totiviridae*	(Garlapati and Wang, 2004)
Gypsy	Drosophila retrotransposon	*Retroviridae*	(Ronfort *et al.*, 2004)
HAV	Hepatitis A virus	*Picornaviridae*	(Brown *et al.*, 1994)
HCV	Hepatitis C virus	*Flaviviridae*	(Tsukiyama-Kohara *et al.*, 1992)
HHV8 / KSHV	Kaposi's sarcoma-associated herpesvirus	*Herpesviridae*	(Bieleski and Talbot, 2001)
HIV-1	Human immunodeficiency virus type 1	*Retroviridae*	(Buck *et al.*, 2001)
HIV-2	Human immunodeficiency virus type 2	*Retroviridae*	(Herbreteau *et al.*, 2005)
HoCV-1	Homalodisca coagulata virus 1	*Dicistroviridae*	(Hunnicutt *et al.*, 2006)
HRV	Human rhinovirus	*Picornaviridae*	(Borman and Jackson, 1992)
HSV	Herpes simplex virus	*Herpesviridae*	(Griffiths and Coen, 2005)

Table 1. (Continued).

Idefix	Drosophila retrotransposon	*Retroviridae*	(Meignin *et al.*, 2003)
LINE-1	Mouse retrotransposon	*Retroviridae*	(Li *et al.*, 2006)
MHV-68	Murine gammaherpesvirus 68	*Herpesviridae*	(Coleman *et al.*, 2003)
MMLV / Mo-MuLV	Moloney murine leukemia virus	*Retroviridae*	(Vagner *et al.*, 1995b)
PEV-8	Porcine enterovirus 8	*Picornaviridae*	(Chard *et al.*, 2006)
PSIV	Plautia stali intestinalis virus	*Dicistroviridae*	(Sasaki and Nakashima, 1999)
PTV-1	Porcine teschovirus 1	*Picornaviridae*	(Kaku *et al.*, 2002)
PV	Poliovirus	*Picornaviridae*	(Dorner *et al.*, 1984; Pelletier and Sonenberg, 1988)
REV-A	Avian reticuloendotheliosis virus type A	*Retroviridae*	(Lopez-Lastra *et al.*, 1997)
RhPV	Rhopalosiphum pali virus	*Dicistroviridae*	(Domier *et al.*, 2000)
RSV	Rous sarcoma virus	*Retroviridae*	(Deffaud and Darlix, 2000)
SIV	Simian immunodeficiency virus	*Retroviridae*	(Ohlmann *et al.*, 2000)
SV2	Simian virus 2	*Picornaviridae*	(Chard *et al.*, 2006)
SV40	Simian virus 40	*Polyomaviridae*	(Yu and Alwine, 2006)
SVDV	Swine vesicular disease virus	*Picornaviridae*	(Sakoda *et al.*, 2001)
TMEV	Theiler's murine encephalomyelitis virus	*Picornaviridae*	(Pilipenko *et al.*, 2000)
TrV	Triatoma virus	*Dicistroviridae*	(Czibener *et al.*, 2005)
TSV	Taura syndrome virus	*Dicistroviridae*	(Hatakeyama *et al.*, 2004)
VL30m	Mouse retrotransposon	*Retroviridae*	(Lopez-Lastra *et al.*, 1999)
WSSV	Shrimp white spot syndrome virus	*Nimaviridae*	(Han and Zhang, 2006)

Table 2. List of plant viruses utilizing a cap-independent translation.

IRES containing virus (abbrev.)	Full name of the virus	Family / Genus	7mG cap / poly(A)	Mechanism	References
BYDV	Barley yellow dwarf virus	*Luteoviridae / Luteovirus*	- / -	TE in 3'-UTR, 18S rRNA complementarity, cap-independent scanning from 5'-end, kissing loops	(Wang *et al.*, 1997; Wang *et al.*, 1999; Guo *et al.*, 2001)

Table 2. (Continued)

crTMV	Crucifer infecting tobamovirus	*Tobamovirus*	+ / -	5' IRES	(Ivanov *et al.*, 1997)
HCRSV	Hibiscus chlorotic ringspot virus	*Tombusviridae*	- / -	5' IRES	(Koh *et al.*, 2003)
PLRV	Potato leafroll polerovirus	*Luteoviridae / Polerovirus*	VPg / -	5' IRES	(Jaag *et al.*, 2003)
RCNMV RNA1	Red clover necrotic mosaic virus	*Tombusviridae*	- / -	TE in 3'-UTR	(Mizumoto *et al.*, 2003)
STNV	Satellite tobacco necrosis virus	*Tombusviridae*	- / -	TE in 3'-UTR, eIF4F and eIFiso4F recruitment	(Gazo *et al.*, 2004)
TBSV	Tomato bushy stunt virus	*Tombusviridae*	- / -	TE in 3'-UTR, cap-independent scanning from 5'-end, kissing loops	(Fabian and White, 2004, , 2006)
TCV 1.45kb sgRNA	Turnip crinkle virus	*Tombusviridae*	- / -	TE in 5'-UTR and 3'-UTR, cap-independent	(Qu and Morris, 2000)
TEV	Tobacco etch potyvirus	*Potyviridae*	VPg / +	TE in 5'-UTR, 18S rRNA complementarity, VPg recruits eIF4E and eIFiso4E, IRES	(Niepel and Gallie, 1999; Schaad *et al.*, 2000; Zeenko and Gallie, 2005)
TMV	Tobacco mosaic virus	*Tobamovirus*	+ / -	TE in 5'-UTR, eIF4F and eIFiso4F recruitment	(Gallie, 2002)
TNV-A sgRNA2	Tobacco necrosis virus A	*Tombusviridae*	- / -	TE in 3'-UTR, 18S rRNA complementarity, kissing loops	(Meulewaeter *et al.*, 2004)
TNV-D	Tobacco necrosis virus D	*Tombusviridae*	- / -	TE in 3'-UTR, kissing loops	(Shen and Miller, 2004)
TuMV	Turnip mosaic potyvirus	*Potyviridae*	VPg / +	5' IRES	(Basso *et al.*, 1994)

Table 3. List of eukaryotic IRESs reported in 5' UTRs of cellular mRNAs (82 in total).

IRES containing gene	Donor organism	References
α-CaM kinase II	*Rattus norvegicus*	(Pinkstaff *et al.*, 2001)
AML1 / RUNX1	*Homo sapiens*	(Pozner *et al.*, 2000)
Antennapedia	*Drosophila melanogaster*	(Oh *et al.*, 1992; Ye *et al.*, 1997)
APAF-1	*Homo sapiens*	(Coldwell *et al.*, 2000)
APC	*Homo sapiens*	(Heppner Goss *et al.*, 2002)
ARC	*Rattus norvegicus*	(Pinkstaff *et al.*, 2001)
AT1R	*Homo sapiens*	(Martin *et al.*, 2003)
BAG-1	*Homo sapiens*	(Packham *et al.*, 1997; Mitchell *et al.*, 2001)
BCL-2	*Homo sapiens*	(van Eden et al., 2004a)
Bcl-xL	*Mus musculus*	(Yoon *et al.*, 2006)
BiP	*Drosophila melanogaster*	(Macejak and Sarnow, 1991)
CAT-1	*Rattus norvegicus*	(Fernandez *et al.*, 2001)
CBFA1 / RUNX2	*Mus musculus*	(Xiao *et al.*, 2003)
Cx26	*Homo sapiens*	(Lahlou *et al.*, 2005)
Cx32	*Mus musculus*	(Hudder and Werner, 2000)
Cx43	*Rattus norvegicus*	(Schiavi *et al.*, 1999)
Cyclin D1	*Homo sapiens*	(Shi *et al.*, 2005)
CYR61	*Homo sapiens*	(Johannes *et al.*, 1999)
c-JUN	*Gallus gallus*	(Sehgal *et al.*, 2000)
c-MYC	*Homo sapiens*	(Nanbru *et al.*, 1997)
DAP5 / p97 / NAT1	*Homo sapiens*	(Henis-Korenblit *et al.*, 2000)
dendrin	*Rattus norvegicus*	(Pinkstaff *et al.*, 2001)
eIF4G	*Homo sapiens*	(Gan and Rhoads, 1996; van Eden *et al.*, 2004a)
E2F6	*Mus musculus*	(Dahme *et al.*, 2002)
ELH	*Aplysia californica*	(Dyer *et al.*, 2003)
FGF1	*Homo sapiens*	(Martineau *et al.*, 2004)
FGF2	*Homo sapiens*	(Vagner *et al.*, 1995a)
FMR1	*Homo sapiens*	(Chiang *et al.*, 2001)
GLI1	*Mus musculus*	(Wang and Rothnagel, 2001)
GRP58	*Homo sapiens*	(Ferguson *et al.*, 1998; Qin and Sarnow, 2004; Scheu *et al.*, 2006)
GluR2	*Rattus norvegicus*	(Myers *et al.*, 2004)
Gtx	*Mus musculus*	(Chappell *et al.*, 2000)
Hairless	*Drosophila melanogaster*	(Chartrand *et al.*, 2002)
HAP4	*Saccharomyces cerevisiae*	(Iizuka *et al.*, 1994; Hecht *et al.*, 2002; Seino *et al.*, 2005)
HIAP2 / c-IAP1 / MIHB / BIRC2	*Homo sapiens*	(Warnakulasuriyarachchi et al., 2003; van Eden *et al.*, 2004a; Warnakulasuriyarachchi *et al.*, 2004)
Hif-1α	*Mus musculus*	(Lang *et al.*, 2002)
HSP70 HSP70	*Drosophila melanogaster* *Homo sapiens*	(Rubtsova *et al.*, 2003; Hernandez *et al.*, 2004)
HSP90	*Drosophila melanogaster*	(Ahmed and Duncan, 2004)
HSP101	*Zea mays*	(Dinkova *et al.*, 2005)

Table 3. (Continued).

IGF-IR	*Rattus norvegicus*	(Fronzes *et al.*, 2003)
IGF-II	*Homo sapiens*	(Pedersen *et al.*, 2002)
KCNA4 / Kv1.4	*Mus musculus*	(Negulescu *et al.*, 1998; Kim *et al.*, 2004)
La1	*Homo sapiens*	(Carter and Sarnow, 2000)
LEF-1	*Homo sapiens*	(Jimenez *et al.*, 2005)
L-MYC	*Homo sapiens*	(Jopling *et al.*, 2004)
MAP2	*Rattus norvegicus*	(Pinkstaff *et al.*, 2001)
MNT	*Homo sapiens*	(Le Quesne *et al.*, 2001)
MS	*Homo sapiens*	(Oltean and Banerjee, 2005)
MYT2	*Homo sapiens*	(Kim *et al.*, 1998)
NAP1L1	*Homo sapiens*	(Qin and Sarnow, 2004)
NDST1 – NDST4	*Mus musculus*	(Grobe and Esko, 2002)
NRF / NF-κB	*Homo sapiens*	(Oumard *et al.*, 2000)
NPM1	*Homo sapiens*	(Qin and Sarnow, 2004)
Nkx6.1	*Mus musculus*	(Watada *et al.*, 2000)
Notch2	*Felix sp.*	(Lauring and Overbaugh, 2000)
n-MYC	*Homo sapiens*	(Jopling and Willis, 2001)
ODC1	*Homo sapiens*	(Pyronnet *et al.*, 2000)
PDGF2 / c-SIS	*Homo sapiens*	(Bernstein *et al.*, 1997; Han *et al.*, 2003)
$p58^{PITSLRE}$	*Homo sapiens*	(Cornelis *et al.*, 2000)
PKCδ	*Rattus norvegicus*	(Morrish and Rumsby, 2002)
PIM-1	*Homo sapiens*	(Johannes *et al.*, 1999; Wang *et al.*, 2005)
$p27^{kip}$	*Homo sapiens*	(Miskimins *et al.*, 2001; Liu *et al.*, 2005)
p53	*Homo sapiens*	(Ray *et al.*, 2006)
RBM3	*Mus musculus*	(Chappell *et al.*, 2001)
RPS18C	*Arabidopsis thaliana*	(Vanderhaeghen *et al.*, 2006)
RC3	*Rattus norvegicus*	(Pinkstaff *et al.*, 2001)
rgr	*Oryctolagus cuniculus*	(Hernandez-Munoz *et al.*, 2003)
Rpr	*Drosophila melanogaster*	(Hernandez *et al.*, 2004)
SCAMPER	*Canis familiaris*	(De Pietri Tonelli *et al.*, 2003)
SMAD5	*Homo sapiens*	(Shiroki *et al.*, 2002)
SNM1	*Homo sapiens*	(Zhang *et al.*, 2002)
spliced leader RNA (SL RNA)	*Leishmania tarentolae*	(Zeiner *et al.*, 2003)
TIF4631	*Saccharomyces cerevisiae*	(Zhou *et al.*, 2001; Altmann *et al.*, 2004; Mauro *et al.*, 2004; Verge *et al.*, 2004)
TrkB	*Homo sapiens*	(Dobson *et al.*, 2005)
Ubx	*Drosophila melanogaster*	(Ye *et al.*, 1997)
UNR	*Homo sapiens*	(Zuidmeer *et al.*, 2005)
URE2	*Saccharomyces cerevisiae*	(Komar *et al.*, 2003)
UtrophinA	*Mus musculus*	(Giersing *et al.*, 2005)
V1bR	*Rattus norvegicus*	(Rabadan-Diehl *et al.*, 2003)
VEGF	*Homo sapiens*	(Huez *et al.*, 1998)
XIAP	*Homo sapiens and Mus musculus*	(Schneider *et al.*, 2001; van Eden *et al.*, 2004a)
YAP1	*Saccharomyces cerevisiae*	(Zhou *et al.*, 2001)

Table 4. List of IRESs synthesized *de novo*.

IRES name	Functional in this host	Mechanism	References
ARC-1	*Oryza sativa*	rRNA complementarity	(Akbergenov *et al.*, 2004)
ICS1-23b	*Rattus norvegicus*	rRNA complementarity	(Owens *et al.*, 2001)
ICS2-17.2	*Rattus norvegicus*	rRNA complementarity	(Owens *et al.*, 2001)
PS3	*Homo sapiens*	rRNA complementarity	(Venkatesan and Dasgupta, 2001)
PS4	*Homo sapiens*	rRNA complementarity	(Venkatesan and Dasgupta, 2001)
14 IRESs	*Saccharomyces cerevisiae*	rRNA complementarity	(Zhou *et al.*, 2003)

Experimental Difficulties Make Proof of IRES a Challenge

The existence of any IRES segment is usually evidenced experimentally by transfer of the putative IRES region from the donor cellular mRNA or viral RNA to reporter RNA. This is very often followed by *in vivo* analyses of the translation levels of the two reporter genes lying in tandem on bicistronic RNA and surrounding inserted putative IRES from both its sides. In most of these experiments the transcription of bicistronic RNA is driven by polymerase II promoters in the cell nucleus and thus the translation of the first cistron is cap-dependent and usually very efficient. Translation of the second cistron is expected to be cap-independent and – depending on the activity of the particular IRES – is sometimes quite low. The protein product of the first (5'-proximal) reporter gene is used both for determining the transfection efficiency and for normalizing of the signal derived from translation of the second cistron. Thus, the translation of the second cistron relies on the putative IRES activity of the sequence inserted between the two reporter genes. This fairly simple assay has many pitfalls. When the difference between the measured translational activities of the first and second cistron is high, the weak activity of the putative IRES can be concealed or mimicked by the accidental occurrence of shorter, functionally monocistronic mRNAs containing solely the second cistron. This situation can happen due to the presence of cryptic splicing sites, breakage hot-spots within the mRNA molecule or due to cryptic transcription initiation. Such transcripts could be capped and thus have a very good chance to be translated as efficiently as the first reporter gene in the master bicistronic mRNA. Therefore, the experiments evaluating the IRES activity are very often prone to produce artificial data and must be accompanied by a number of precisely designed controls. On the other hand, it should be noted that the presence of cryptic promoters, cryptic splicing sites, *etc.* within the IRES segment tested does not necessarily mean failure in proving the IRES activity. Such an IRES segment can pass other assays based for example on direct RNA transfection into the cells, on the use of 7mG cap analogs or on depletion or modification of cap-binding translation initiation factors.

The methodology used to determine IRES segments has been extensively discussed during the past few years (Jackson, 2000; Kozak, 2001; Schneider *et al.*, 2001; van Eden *et al.*, 2004a; Kozak, 2005) and some of the former conclusions have been re-evaluated. Besides the already mentioned IRESs containing cryptic promoters, care must be taken in selecting the proper plasmid backbone. Several internal ribosome entry sites (Bcl2 (Sherrill *et al.*, 2004), c-IAP (van Eden *et al.*, 2004b), Kv1.4 (Jang *et al.*, 2004), L-myc (Jopling *et al.*, 2004), CVB3 (Jang *et al.*, 2004) have been reported to reveal a cryptic splicing site in the context of the chimeric intron containing pRF vector and its derivatives (Stoneley *et al.*, 1998; Jang *et al.*, 2004). In spite of the unknown nature of the function of most of the reported cellular IRESs it is possible to classify them into the following four simple groups:

1. The IRES region shows an apparent complementarity to rRNA which is expected to confer the IRES activity observed (Chappell *et al.*, 2000). Such IRES functionally resembles the prokaryotic Shine-Dalgarno sequence.
2. The IRES activity is mediated by protein-RNA or protein-protein interactions. Either direct ribosome binding or the action of various *trans*-acting protein factors (ITAFs) is responsible for the IRES dependent translation initiation. Currently, the research in this field is in the stage of accumulating new descriptive data about the existence of novel ITAFs and their specificities.
3. The mechanism of IRES action has not been found or even studied. Most of the cellular IRESs belong to this group.
4. Those putative IRES segments whose characterization has been reported without providing proper controls or have already been claimed as experimental artifacts.

IRESITE PRESENTS BOTH STRUCTURAL AND EXPERIMENTAL DATA

The IRESite database is dedicated to presenting exhaustive information about the particular internal ribosome entry sites including their structural characteristics, interacting partners as well as experimental evidence of IRES function (Mokrejš *et al.*, 2006). In principle it contains two different groups of records covering either the natural properties of the RNA molecule containing a particular IRES or describing the experiments, which were performed to obtain an evidence or further data of the IRES function. The first group of the database entries is denominated as *natural*, the latter is referred to as *engineered*. IRESite contains a powerful search engine and, where applicable, the individual entries are linked to PubMed, GenBank and other publicly available resources.

The aim of the *natural* entries is to summarize available data about IRESs in their natural sequence context including the annotated full-length sequences of mRNAs or viral RNAs, IRES lengths and positions, IRES secondary structures when available and the reported interacting proteins.

In contrast, the *engineered* records, while reflecting variants in the experimental setup, store information about the results of IRES translation experiments carried both *in vitro* and *in vivo*. These records accommodate data about the utilized plasmids, the regulatory sequences, the reporter genes, observed IRES activity and much more. Basically, the *engineered* records may also present sequences of, and experimental data about various mutated forms of particular IRES, which are frequently employed for detailed analyses of IRES function. A more detailed description of the IRESite database has been published recently elsewhere (Mokrejš *et al.*, 2006).

One of the most important points, which make the IRESite database unique, is the declared aim to provide and annotate the full-length RNA sequences both for natural and engineered records whenever possible. However, this is not a simple task. In most cases such sequences are not available in public sequence repositories or have never been even published. Therefore either the authentic sequence data have to be obtained from original authors or the sequences have to be reconstituted *in silico* from the most similar sequences available in GenBank, printed publications and other resources.

Frequently, new DNA vectors are updated for the particular research from preexisting ones, which were originally created for other purposes. Sometimes, the precise sequence analysis, which is necessary for detailed annotation of IRESite entries and could not be accomplished without the kind help of the original authors, reveals hidden properties of the provided respective plasmids. This situation can be exemplified by the pRSTF-CBV3 plasmid containing human coxsackie virus IRES (Jang *et al.*, 2004) (kindly provided by J. Jimenez). We have found that the second reporter gene coding for the firefly luciferase is fused in frame at its N-terminus with the 2A peptide encoding the sequence from the foot-and-mouth disease virus (Figure 1). This "protease" peptide does not cleave the fusion protein but rather acts as a modulator of the ribosomal activity and causes release of the N-terminal peptide from the ribosome, while the ribosome continues in the translation of the rest of the mRNA template, starting the following protein with upcoming prolyl-tRNA. However, this "cleavage" action of 2A peptide is not hundred percent efficient and a part of the ribosomes stop further translation (Donnelly *et al.*, 2001a; Donnelly *et al.*, 2001b). Although the yield of the second reporter product and thus the measured activity of CBV3 IRES could be perhaps underscored, in this case we expect no important influence on the quality of the obtained results.

Figure 1. Part of the pRSTF-CBV3 vector sequence. The coding regions are depicted in uppercase letters, the *Renilla* and firefly luciferase genes are coloured in green, IRES from the human coxsackie virus B3 is shown in blue, stem-loop structure used to block ribosomal reading-through from the first cistron is in yellow and the FMDV 2A peptide coding sequence is coloured in magenta.

In summary, IRESite can provide useful information not only about a particular IRES itself but also about the surrounding sequences both in the coding and non-coding RNA regions. The growing number of *engineered* records in IRESite facilitates a comparison of IRES translation efficiencies as well as an evaluation of the different experimental setups employed for IRES characterization.

By September 2006 IRESite covered 56 unique *natural* RNAs of cellular or viral origin (Table 5). In some cases multiple, either non-contiguous or overlapping sequence regions of the same transcript have been shown to possess the IRES activity and therefore such IRESs span multiple lines in the table. In addition the database contains 101 *engineered* records with annotated key experimental parameters and 25 records containing sequences of various control plasmids.

What Can we Gather from the Size and the GC Content of IRES Segments?

We analyzed and correlated the sizes, the number of AUG codons and the GC content of the IRES segments and corresponding full-length RNA transcripts and viral RNAs, which are currently included in the IRESite database (Table 5).

Figure 2 and Figure 3 show the analyses of the IRES GC contents plotted against the mRNAs GC contents and the IRES lengths, respectively. Both analyses demonstrate that the group of cellular IRESs is more heterogeneous than the group of viral IRESs. The GC content of viral IRES segments is generally only a little bit higher than the GC content of the corresponding viral +RNAs. Contrary to the common opinion, the selection pressure for increased GC content and thus for increased occurrence of secondary and higher order IRES structures will not be probably much different from the other parts of the viral genomic +RNA. Figure 2 shows that while viral IRESs create a homogeneous group, cellular IRES segments appear to separate into the three distinct clusters. One group consists of short or medium sized IRESs with very high GC content, the second involves IRESs of the same size as the first group but with low or intermediate GC content and the third group contains extremely long IRES segments. The biological relevance of this finding is unclear, it can reflect both the biological reality as well as the biased "selection by scientist" driven, instead of the power of natural selection, by the opinion that the internal ribosome entry sites of cellular mRNAs more frequently occur in long or highly structured 5' UTRs (Gan and Rhoads, 1996; Huez *et al.*, 1998; Negulescu *et al.*, 1998; Oumard *et al.*, 2000; Pozner *et al.*, 2000; Watada *et al.*, 2000; Grobe and Esko, 2002; De Pietri Tonelli *et al.*, 2003; Sherrill *et al.*, 2004; van Eden *et al.*, 2004b; Jimenez *et al.*, 2005).

Analysis of the number of the AUG codons in the untranslated part of the internal ribosome entry sites (Figure 4 and 5) again clearly shows the apparent uniformity of viral IRESs, which generally have the number of the AUG codons roughly corresponding to the expected frequency of the AUG codon in random nucleotide sequences. Figure 5 presents the number of the AUG codons per hundred nucleotides of the IRES plotted against the corresponding IRES GC content. With one exception, this analysis revealed that only the AT-rich cellular IRES segments appear above the line of the theoretically expected AUG frequency in the random nucleotide sequence. Additionaly, most of the cellular IRESs with higher GC content contain fewer than predicted or no AUG codons in their untranslated regions. On the contrary, the majority of viral IRESs is grouped around the line of the theoretically expected AUG frequency regardless of their GC content.

The fruit fly retrotransposonal IRESs form a distinct group in all analyses. We cannot explain now if the reason for such a difference consists in special features of the retrotransposons themselves or if it is rather caused by the long period of their evolution within the fruit fly genomic environment. Interestingly, the IRES from the HIV retrovirus contains no AUG and thus resembles retrotransposons in this respect. Five cellular IRESs involved in the present study have been reported to contain promoter in their cDNA. These IRESs are marked with open circle and apparently do not form a distinguishable group in any of the graphs.

If we suppose that, according to the cap-dependent scanning model of translation initiation, the translatability of particular mRNA is lowered by secondary structures (high GC content) or multiple AUG containing 5' UTRs (Kozak, 1989), then we can hypothesize that the presented results confirm that the 5' UTRs containing cellular IRESs but no viral IRESs are at least on some occasions also translated by the scanning mechanism.

Table 5. Cellular and viral IRES segments covered by IRESite by September 2006.

IRESite_Id	gene name	mRNA length	IRES name	IRES length	IRES origin
57	AML1/RUNX1	7288	AML1/RUNX1	1561	cellular
71	Antp	3490	Antp-CDE	1730	cellular
71	Antp	3490	Antp-D	252	cellular
71	Antp	3490	Antp-DE	408	cellular
110	Apaf-1	7042	Apaf-1	233	cellular
103	BCL2	5086	BCL2	1137	cellular
108	c-IAP1	3753	c-IAP1	1159	cellular
48	c-jun	1717	c-jun	301	cellular
35	c-myc	1958	c-myc	395	cellular
117	DAP5	3820	DAP5	306	cellular
122	FMR1	4397	FMR1	252	cellular
204	Gtx	1244	Gtx-1-120	120	cellular
204	Gtx	1244	Gtx-1-166	166	cellular
204	Gtx	1244	Gtx-1-196	196	cellular
204	Gtx	1244	Gtx-133-141	9	cellular
66	Hairless	5621	hairless	435	cellular
116	hap4	2937	Sc_HAP4	272	cellular
83	hAT1R-A	2263	AT1R_var2	272	cellular
82	hAT1R-B	2347	AT1R_var1	356	cellular
84	hAT1R-C	2321	AT1R_var3	330	cellular
85	hAT1R-D	2405	AT1R_var4	414	cellular
1	Hif1a	3973	HIF1a	257	cellular
58	hSNM1	4468	hSNM1	918	cellular
67	Hsp70Aa	2630	hsp70	503	cellular
118	HSPA1A	2404	hsp70	195	cellular
124	Kcna4	4780	Kcna4	1197	cellular

Table 5. (Continued)

89	L-myc	3258	L-myc	52	cellular
220	LEF1	3612	LEF1	1167	cellular
51	MNT	4865	MNT	193	cellular
49	MYT2	2308	MYT2	156	cellular
55	n-myc	2613	n-myc	320	cellular
107	odc1	2035	ODC1	302	cellular
138	PIM1	2717	Pim-1	396	cellular
8	Rbm3	1430	Rbm3	22	cellular
62	Rpr	851	reaper	168	cellular
111	Scamper	1243	Scamper	365	cellular
65	Ubx	4729	ultrabithorax	968	cellular
65	Ubx	4729	ultrabithorax	594	cellular
81	UNR	4115	UNR	429	cellular
115	Ure2	1069	Sc_Ure2	165	cellular
109	XIAP	8751	XIAP	460	cellular
37	BVDV1	12573	BVDV1_1-385	385	viral
37	BVDV1	12573	BVDV1_29-391	363	viral
40	CrPV	9185	CrPV_5NCR	708	viral
40	CrPV	9185	CrPV_IGR	192	viral
39	crTMV	6298	crTMV_IRES_CP	148	viral
148	CSFV	12311	CSFV	373	viral
225	CVB3	7399	CVB3	750	viral
140	EMCV	7835	EMCV	576	viral
29	ERAV_1	7697	ERAV_1	714	viral
26	ERBV_1	8828	ERBV_1	718	viral
54	GBV-B	9399	GBV_B	424	viral
38	GBV_A	9653	GBV_A	693	viral
41	GBV_C	9378	GBV_C	630	viral
69	Gypsy	7468	gypsy1	330	viral
69	Gypsy	7468	gypsy2	261	viral
69	Gypsy	7468	gypsyD1	517	viral
42	HAV	7478	HAV_HM175	584	viral
222	HCV_type_1b	9416	HCV_type_1b	341	viral
73	HIV-1	9181	HIV-1	233	viral
139	HRV	7108	HRV-2	614	viral
68	Idefix	6919	idefix	521	viral
59	PSIV	8797	PSIV_IGR	188	viral
77	TSV	10205	TSV_IGR	250	viral

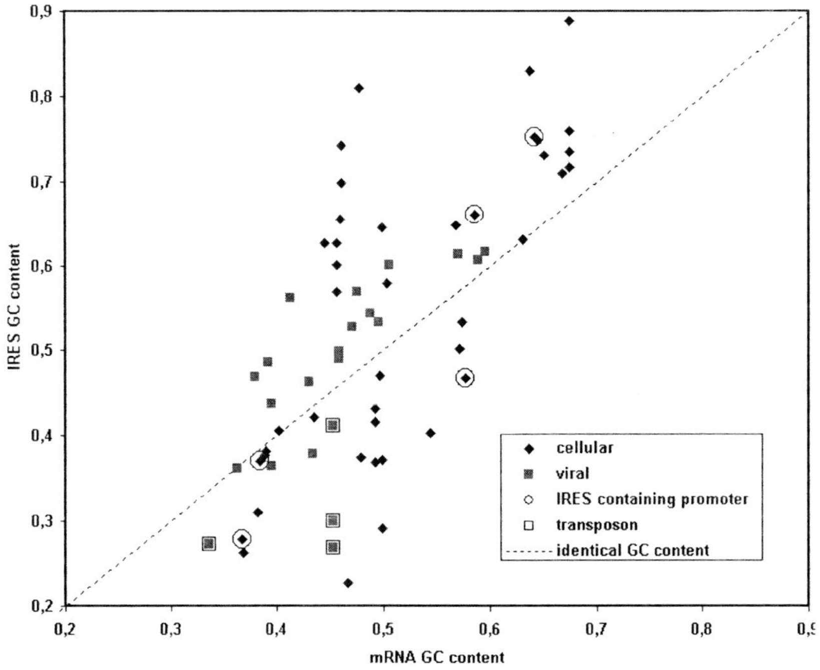

Figure 2. Comparison of the GC content within IRES and the respective mRNA or viral RNA. 3'-UTRs including possible poly(A) sequences were excluded from the analysis. Retrotransposonal IRESs are distinguished from viral ones by open squares. Cellular IRESs containing promoter in their cDNA are marked with circles.

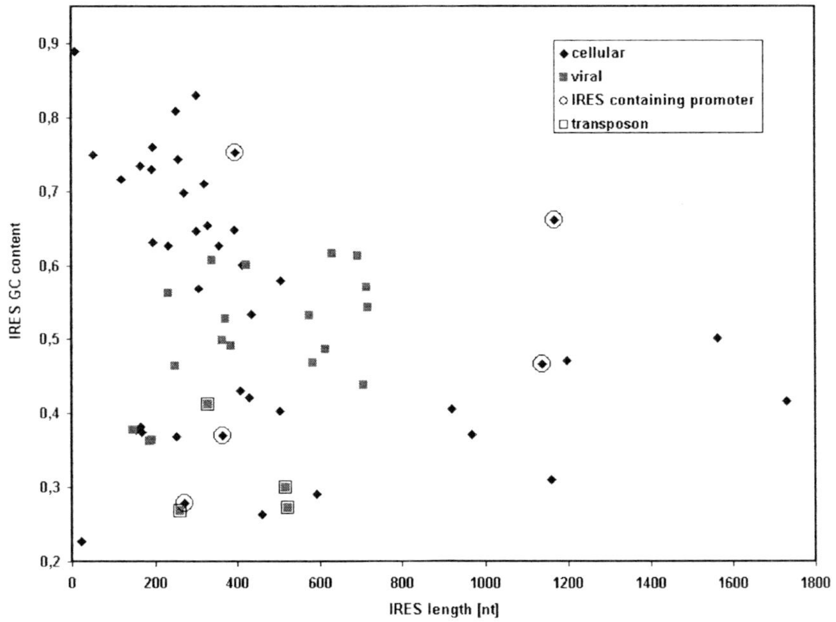

Figure 3. Correlation of the IRES GC content and its length. Retrotransposonal IRESs are distinguished from viral ones by open squares. Cellular IRESs containing a promoter in their cDNA are marked with circles.

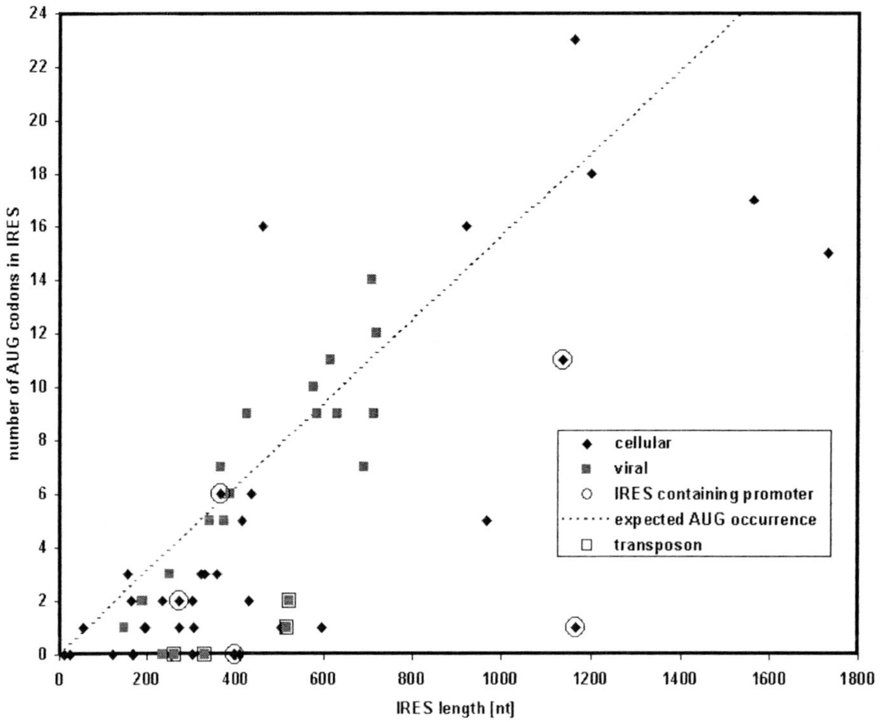

Figure 4. The number of AUG codons within the IRES sequences correlated to the IRES lengths. Retrotransposonal IRESs are distinguished from viral ones by open squares. Cellular IRESs containing a promoter in their cDNA are marked with circles.

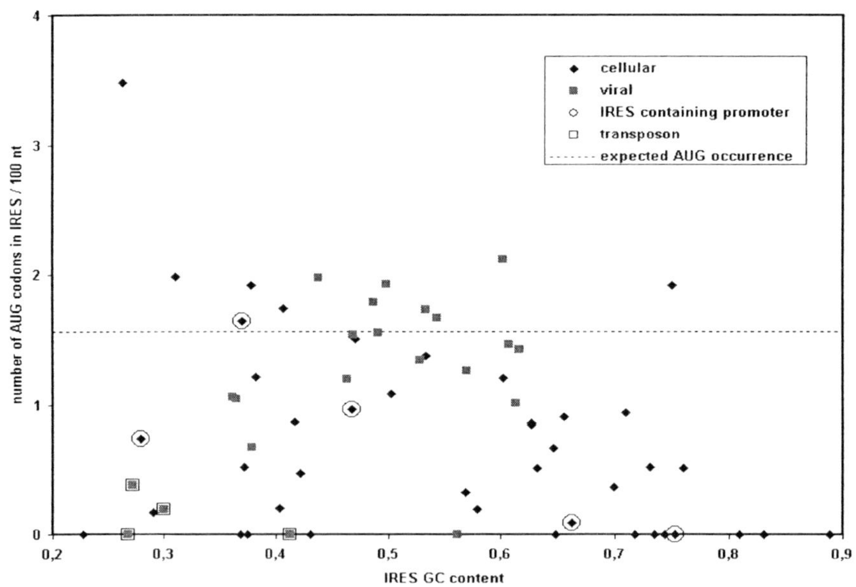

Figure 5. The frequency of the AUG codons within the IRES sequences versus IRES GC content. Retrotransposonal IRESs are distinguished from viral ones by open squares. Cellular IRESs contaning promoter in their cDNA are marked with circles.

IRESITE IN THE CONTEXT OF OTHER RNA-ORIENTED SEQUENCE DATABASES

A number of databases cover various EST and cDNA sequences, which are frequently incompletely sequenced mRNAs and the chances to analyze complete 5'-UTR regions containing IRESs are therefore limited. Besides IRESite, there are two other databases (UTRdb and IRESdb) dealing to some extent with internal ribosome entry sites.

UTRDB CONTAINS IRESS AUTOMATICALLY PREDICTED BY SEQUENCE PATTERN MATCHES

The UTRdb database (Pesole *et al.*, 1998) (`http://www.ba.itb.cnr.it/UTR/`) is automatically generated by a UTRdb_gen program which parses input EMBL database records and extracts from them the 5'-UTR sequences. Not all EMBL records have tagged the 5'-UTR regions but the UTRdb_gen program uses its own heuristics to find 5' and 3'-UTR based on the position of coding sequences, exons, introns and other feature keys although the details have never been disclosed. Database flat files are available for download from the FTP server directory at `ftp://bighost.ba.itb.cnr.it/pub/Embnet/Database/UTR/data` and are regularly updated with newer EMBL releases. A pattern-matching algorithm is used to find known sequence motifs in UTRdb records. To our concern, the prediction of IRES segments in UTRdb is based on the presence of a predicted Y-type shaped stem-loop structure followed by a stem-loop just upstream of the AUG initiation codon (Le and Maizel, 1997). The model is based on analysis of mRNAs coding for the human immunoglobulin heavy chain binding protein (BiP), fruit fly Antennapedia (Antp) and the human fibroblast growth factor 2 (FGF-2).

Figure 6 shows a full text definition of the IRES element in the UTRSite database including the search pattern (ID U0015). The UTRSite database contains currently 52 such records (in the downloadable file) whereas only 40 of them are accessible through the UTRSite Sequence Retrieval System (SRS). Users of the UTRSite web server can submit their own sequence and the server searches through the query for the recognized sequence pattern of 52 (or 40?) motifs.

Users can also search UTRdb for sequences similar to their query by the BLAST algorithm and SRS tools (text search, feature search, *etc.*). The SRS search for the feature key 'IRES' returns 30 168 current entries from the non-redundant UTR database. Interestingly, the search against the redundant database which should contain more entries returns only 14 491 entries. It seems that the redundant and non-redundant files have swapped names. This suspicion was confirmed by the downloaded flat file data from the FTP server. In this text we will refer to the files as redundant or non-redundant based on their actual contents without regards to the actual file names (redundant files have mistakenly `_nr.dat` in their filename whereas the non-redundant files are labeled `_r.dat`). UTRdb is not an absolutely non-redundant database because certain overlaps between sequences as well as pairwise sequence similarity up to 95% (Pesole *et al.*, 1998) are allowed between any two records.

However, the large number of IRES hits in both datasets still cannot be easily explained and was subjected to our further studies.

```
ID:               U0015; date: 08-05-2000
Name:             Internal Ribosome Entry Site (IRES)
Pattern:          r1={au,ua,cg,gc,gu,ug}
                  p1=5...6 0...6 p2=5...6 p3=0...2
                  p4=5...8 p5=3...8
                  r1~p4[1,0,0] p6=5...8 p7=3...5
                  r1~p6[1,0,0] p8=0...5
                  r1~p2[1,0,0] 0...6
                  r1~p1[1,0,0] 2...5 p9=5...6 p10=3...8
                  r1~p9[1,0,0] 3...10 ATG 97...97 $
Taxon_Range:      5UTR, 5'UTRs
Description:      Internal mRNA ribosome binding is a mechanism of
                  translation initiation alternative to the
                  conventional 5'-cap dependendent ribosome scanning
                  mechanism.
                  The internal initiation mechanism, first observed
                  in picornavirus, is also found in some cellular mRNAs,
                  such as human immunoglobulin heavy chain binding protein
                  (BiP) mRNA, Antennapedia mRNA of drosophila and FGF-2 mRNA.
                  This mechanism may be advantageous for specific translation
                  of mRNAs during cell cycle when underphosphorilated eIF-4F
                  prevents conventional translation initiation. The internal
                  initiation mechanism is dependent on a cis-acting element,
                  defined as internal ribosome entry site (IRES), located
                  in the 5'UTR region of the mRNA.
                  Comparative analysis of known IRES in some cellular mRNAs
                  identified a common structural motif forming a Y-type stem-loop
                  structure followed by the AUG triplet or followed by additional
                  stem-loop structures and the AUG triplet.
Bibliography:     [1]
   Authors        Le, S.Y., and Maizel,J.V.,Jr. (1997)
   Title          "A common RNA structural motif involved in the internal initiation of translation
   Title          of cellular mRNAs"
   Journal        Nucleic Acids Res.25,362-69
//
```

Figure 6. Annotation of the IRES segment (U0015) including its searching pattern in the UTRSite database.

A detailed examination of the flat files of the UTRdb release 22 (derived from the EMBL release 86 and from the UTRef release 7 data derived from the REFSEQ release 17) confirmed a high number of the retrieved IRESs from UTRdb website (Table 6). To evaluate capability of the Le and Maizel pattern used in UTRdb for prediction of IRES segments (Le and Maizel, 1997) we compared a list of genes predicted to contain IRES segment with the list of already published IRESs. While inspecting the actual records with predicted IRESs we found that at least two types of redundancies increase the obtained counts.

- First, several IRESs can stem from multiple mRNA transcripts derived from a single EMBL gene record. Such transcripts can differ either due to alternative transcription starts or splicing events (while the IRES is predicted in a common exon).
- The second type of redundancy comes from multiple allelic variants of the same gene/mRNA in EMBL database, which somehow leaked through the similarity filter.

To avoid these multiplicated instances of the same gene we decided to pick up every gene name only once. We also eliminated records containing putative chromosomal coding regions, where no gene names have been annotated or no experimental evidence for the corresponding transcript was supposed.

Cellular IRESs are mostly studied in mammals. When we eliminated by gene name IRES containing records common to both human-related files, 1266 entries remained as unique to the non-redundant file (`5UTR.Hum_r.dat`) whereas the redundant file (`5UTR.Hum_nr.dat`) had 2493 IRES containing genes (which do not have a match in the non-redundant file `5UTR.Hum_r.dat`). Thus, the non-redundant and redundant files still do differ significantly and we explain this difference by alternatively used gene names (aliases) for the *de facto* same records and by generally inconsistent annotation used in the projects which deposited their data into public sequence databanks. The appearance of gene names like "`p53 related protein`" or "`GLYCOGEN SYNTHASE KINASE 3 BETA`" demonstrates that the abbreviated gene consists of multiple words. Such records have also been eliminated.

For the purpose of the presented comparative analysis we also eliminated records corresponding to the putative and hypothetical genes by matching certain strings of characters in them because such genes are unlikely to be included among the published IRESs. Examples of the eliminated records are `DKFZP566I1024` (see `UTRdb:BB251405` in `5UTR.Hum_r.dat`), `LOC285888`, `TMP_LOCUS_28`, `C9ORF125`, *etc.*

More thorough analysis by all-against-all BLAST search between all records of each dataset would have to be done to identify all the duplicates. However, even the presented results raised some important issues. We argue it would be more convenient for the user if UTRdb would avoid the problem with gene name aliases, redundancy and hopefully even inconsistent annotation of gene names altogether and predict the recognized 52 (or 40?) features in data from non-redundant, gene-oriented databases like UniGene (Wheeler *et al.*, 2004). Further, although UTRdb is a collection of all 5'-UTRs we see no reason for predicting IRES segments in chloroplast mRNA (`UTRdb:BB407559` derived from `EMBL:AB087484` in file `5UTR.Pln_nr.dat`). We have also found that dsRNA from a chloroplast (`UTRdb:BB407233` derived from `EMBL::AB070653` in the same file) is misplaced with plant genes in this file while we prefer to include it in the file with viral sequences (although this is probably an annotation issue of the upstream databases). Finally, although the UTRdb_gen program correctly parsed EMBL records and determined the 5'-UTR region the provided resulting flat files do not comply with the definition of EMBL file format. At least 5 types of errors in the format were uncovered when we have parsed UTRdb flat file data. A fix for some of the supposedly more common while non-critical errors in the format has been committed to the source code of bioperl in Sept 2006 (Stajich *et al.*, 2002). The fixed EMBL and GenBank formatted flat files used throughout this analysis together with the resulting list of predicted IRES-containing genes are available from http://www.iresite.org/UTRdb.

Results of our examination of UTRdb can be exemplified on its fungal dataset. This analysis showed that the number of fungal entries with the annotated IRES feature in the redundant 5UTR.Fun_nr.dat file reaches 343 whereas only 79 records can be found in the non-redundant 5UTR.Fun_r.dat file. The number of records decreased substantially to 134 and 37 respectively after the cleanup procedure described above.

IRESs Predicted in `5UTR.Fun_nr.dat` After Elimination of Duplicates and Hypothetical and Putative Genes

At least 25 IRESs were predicted in *Saccharomyces cerevisiae* while none of them has been reported to contain IRES yet: ARV1, CKI1, DBM1, HAP2, HEM1, HIS7, LEU3, MSS51, NCE2, NHP6A, OPY1, PDR5, POX1, RAD3, RPL37A, SIN3, SPE1, SPO1, STE14, TPK1, URA6, UTR2, VAS1, WSC3, YAP2.

Similarly, at least 29 IRESs were predicted in *Schizosaccharomyces pombe* while again none of them has been reported to contain IRES: ABC1SP, CDS1, CSH3, CUT17, DFR1, DMF1, FRP1, GHT2, HIS5, HSP70, KRP1, PAS1, PDS5, PHP3, PKA1, PLA1, PTC1, PUB1, RAD9, RES1, RHP55, RPB4, RPB7, SPAPN1, SPC1, SPJ1, STM1, TAP1, UVDE.

IRESs Predicted in `5UTR.Fun_r.dat` after Elimination of Duplicates and Hypothetical and Putative Genes

Only 10 IRESs came out from the previously mentioned 25 genes in the case of *Saccharomyces cerevisiae*: HAP2, HEM1, LEU3, MSS51, PDR5, POX1, SIN3, SPE1, SPO1, VAS1. We speculate that the genes missing in the list disappeared during the automated procedure used to create the non-redundant dataset. Into the resulting set it possibly randomly picked up another record similar in sequence but either lacking abbreviated gene name or containing the abbreviated gene name consisted of multiple words or although the record was correctly annotated the IRES feature was not predicted by pattern match.

Similarly, only 20 IRESs came out from the list of 29 genes listed above for *Schizosaccharomyces pombe*: CSH3, DFR1, DMF1, KRP1, PAS1, PDS5, PHP3, PKA1, PLA1, PTC1, PUB1, RES1, RHP55, RPB7, SCT1, SPAPN1, SPC1, SPJ1, STM1, TAP1.

Similar analysis of human redundant dataset (`5UTR.Hum_nr.dat`) revealed 3477 genes predicted to contain IRES after elimination of duplicates and hypothetical and putative genes. However, only 12 of them have been reported to contain IRES so far in either human, mouse, rat, dog or cat genomes (Table 3): AML1, ARC, CYR61, E2F6, FGF1, FGF2, HSP70, NAT1, NDST3, NPM1, ODC1, SMAD5. 65 of the currently reported mammalian IRESs have not been recognized showing thus that the prediction quality of the pattern search (Le and Maizel, 1997) employed by UTRdb is only 12%. We did not attempt to evaluate whether the predicted IRES regions actually overlap with the published IRES thus the predictions might still include some false positives.

In summary, if the intent of any UTRdb user is to get rather complete albeit redundant list of any genes one should preferably use the redundant files with `_nr.dat` in their names (unless UTRdb curators swap the filenames) which contain many highly similar entries and some genes possibly appear under alternative names several times. The files with `_r.dat` in their names contain the non-redundant set of sequences and one should be aware that the employed automated procedure probably eliminates well annotated genes containing putative IRESs (or any other UTRSite feature as well) in some cases while retaining records with poor annotation with a merely same sequence or with alternative gene name annotated.

Table 6. The number of records predicted in UTRdb to contain IRES is unexpectedly large.

filename	# of IRES containing records	# of IRESs	filename	# of IRES containing records	# of IRESs
5UTR.Fun_nr.dat	343	134	5UTRef.Inv.dat	4,971	1,777
5UTR.Fun_r.dat	79	37	5UTRef.Mam.dat	6,617	673
5UTR.Hum_nr.dat	7,546	3,477	5UTRef.Pln.dat	3,862	1,279
5UTR.Hum_r.dat	5,176	2,297	5UTRef.Pri.dat	8,053	3,804
5UTR.Inv_nr.dat	3,048	1,900	5UTRef.Rod.dat	8,740	5,313
5UTR.Inv_r.dat	441	266	5UTRef.Vrt.dat	4,932	2,367
			SUM of IRESs in REFSEQ	**37,175**	**15,213**
5UTR.Mam_nr.dat	1,983	568			
5UTR.Mam_r.dat	242	97			
5UTR.Mus_nr.dat	4,011	2,782			
5UTR.Mus_r.dat	1,606	977			
5UTR.Patent_nr.dat	80	4			
5UTR.Patent_r.dat	26	1			
5UTR.Pln_nr.dat	5,540	3,617			
5UTR.Pln_r.dat	2,081	1,308			
5UTR.Rod_nr.dat	1,696	1,031			
5UTR.Rod_r.dat	342	172			
5UTR.Vrl_nr.dat	903	125			
5UTR.Vrl_r.dat	3,802	260			
5UTR.Vrt_nr.dat	5,018	2,987			
5UTR.Vrt_r.dat	696	395			
SUM of IRESs in EMBL	**44,659**	**22,435**			

The files derived from the EMBL (filenames starting with `5UTR` are in the left half of the table) and from the REFSEQ (filenames starting with `5UTRef` are in the right half of the table) databases include hypothetical, conserved hypothetical and putative genes (having yet no gene names assigned). The filenames in 1st and 4th column are the names of files available on the UTRdb ftp server and each of them contains a subset of the taxonomically distinct records. The numbers in the 2nd and 5th column represent all predicted IRES containing genes in the respective taxonomic group. The numbers in the 3rd and 6th column represent all predicted IRES containing genes in the respective taxonomic group purified of the hypothetical and the putative genes. Please note that the "non-redundant" and "redundant files" are probably swapped (see the main text for more information). The underlined numbers of IRESs predicted for the human and the mouse indicate the predicted number of IRES containing genes per organism. However, some genes may appear under multiple aliases, making thus the number overrepresented to unknown extent. Files with the following substrings within their filenames contain the nucleotide sequences as indicated: `Fun` – fungal/yeast sequences, `Hum` – human sequences, `Inv` – invertebrate sequences, `Mam` – mammalian sequences except the human and the mouse, `Mus` – mouse sequences, `Patent` – various patented sequences, `Pln` – plant sequences, `Rod` – rodents, `Vrl` – viral sequences and `Vrt` – vertebrates except the human and the mouse.

I.R.E.S. Database (IRESdb) Compiles References to GenBank Records and Literature

IRESdb database (Bonnal *et al.*, 2003) is a manually curated database (http://www.rangueil.inserm.fr/IRESdatabase/) which contains a list of IRES segments of cellular (51 records) and viral (31 records) origin. The website presents IRES segments sorted according to their origin (viral or cellular), according to IRES *trans*-acting factors (ITAFs) interacting with them and according to the function of the proteins encoded by the IRES harboring mRNAs. Thus, it is possible to reach the same IRES through multiple menu items. IRESdb sequence entries consist either both from the annotated 5'-UTRs and the corresponding GenBank link or from the link to the GenBank only. The 5'-UTR sequences are mostly routinely extracted from the respective GenBank entries and are presented just until the initiation AUG codon. A search tool allows the search for text strings through the website by employing probably some external search engine (like Google site search).

IRESdb and IRESite databases both present valuable data and well complement each other. The IRESdb currently presents more data about the ITAFs and enables one to browse the cellular IRESs according to the function of the corresponding mRNAs. In contrast, the IRESite database is focused primarily on annotation of the original full-length RNA sequences usually obtained directly from their authors or after their manual and exigent reconstruction from various electronic and printed resources. IRESite presents the IRES regions conveniently within the full length RNAs. The 5' and 3' borders of the respective IRES can thus be easily described even in the cases when IRES overlaps with the downstream coding region or the particular RNA contains both multiple or modular IRESs. In addition, IRESite annotates the experimental procedures which were used to characterize the particular IRES segment and can even record the IRES secondary structures if available. Both the IRES oriented databases provide links to relevant PubMed literature.

Acknowledgement

We would like to thank all the colleagues who helped us and cooperated with us in building the IRESite database and who kindly provided us with the sequences and their experimental data. We also would like to thank K. Sigler for his language assistance. This work was supported by Czech Grant Agency (Grants No. 204/03/1487 and 301/07/0607), by the Grant Agency of the Charles University (Grant No. 251/2004/B-BIO/PrF) and by the Ministry of Education (Grants No. MSM 0021620813; 1M06014 and LC06066).

References

Ahmed R, Duncan RF. 2004. Translational regulation of Hsp90 mRNA. AUG-proximal 5'-untranslated region elements essential for preferential heat shock translation. *J Biol Chem 279*:49919-49930.

Akbergenov R, Zhanybekova S, Kryldakov RV, Zhigailov A, Polimbetova NS, Hohn T, Iskakov BK. 2004. ARC-1, a sequence element complementary to an internal 18S rRNA segment, enhances translation efficiency in plants when present in the leader or intercistronic region of mRNAs. *Nucleic Acids Res 32*:239-247.

Altmann M, Trachsel H, Vonlanthen M, Vergé V. 2004. Answer to "Reevaluation of the conclusion that IRES-activity reported within the 5′ leader of the *TIF4631* gene is due to promotor activity". *RNA 10*:898.

Bailly JL, Borman AM, Peigue-Lafeuille H, Kean KM. 1996. Natural isolates of ECHO virus type 25 with extensive variations in IRES sequences and different translational efficiencies. *Virology 215*:83-96.

Basso J, Dallaire P, Charest PJ, Devantier Y, Laliberte JF. 1994. Evidence for an internal ribosome entry site within the 5' non-translated region of turnip mosaic potyvirus RNA. *J Gen Virol 75 (Pt 11)*:3157-3165.

Beaulieux F, Zreik Y, Deleage C, Sauvinet V, Legay V, Giraudon P, Kean KM, Lina B. 2005. Cumulative mutations in the genome of Echovirus 6 during establishment of a chronic infection in precursors of glial cells. *Virus Genes 30*:103-112.

Belsham GJ, Brangwyn JK. 1990. A region of the 5' noncoding region of foot-and-mouth disease virus RNA directs efficient internal initiation of protein synthesis within cells: involvement with the role of L protease in translational control. *J Virol 64*:5389-5395.

Berlioz C, Darlix JL. 1995. An internal ribosomal entry mechanism promotes translation of murine leukemia virus gag polyprotein precursors. *J Virol 69*:2214-2222.

Bernstein J, Sella O, Le SY, Elroy-Stein O. 1997. PDGF2/c-sis mRNA leader contains a differentiation-linked internal ribosomal entry site (D-IRES). *J Biol Chem 272*:9356-9362.

Bieleski L, Talbot SJ. 2001. Kaposi's sarcoma-associated herpesvirus vCyclin open reading frame contains an internal ribosome entry site. *J Virol 75*:1864-1869.

Bonnal S, Boutonnet C, Prado-Lourenco L, Vagner S. 2003. IRESdb: the Internal Ribosome Entry Site database. *Nucleic Acids Res 31*:427-428.

Borman A, Jackson RJ. 1992. Initiation of translation of human rhinovirus RNA: mapping the internal ribosome entry site. *Virology 188*:685-696.

Brown EA, Zajac AJ, Lemon SM. 1994. In vitro characterization of an internal ribosomal entry site (IRES) present within the 5' nontranslated region of hepatitis A virus RNA: comparison with the IRES of encephalomyocarditis virus. *J Virol 68*:1066-1074.

Buck CB, Shen X, Egan MA, Pierson TC, Walker CM, Siliciano RF. 2001. The human immunodeficiency virus type 1 gag gene encodes an internal ribosome entry site. *J Virol 75*:181-191.

Carter MS, Sarnow P. 2000. Distinct mRNAs that encode La autoantigen are differentially expressed and contain internal ribosome entry sites. *J Biol Chem 275*:28301-28307.

Coldwell MJ, Mitchell SA, Stoneley M, MacFarlane M, Willis AE. 2000. Initiation of Apaf-1 translation by internal ribosome entry. *Oncogene 19*:899-905.

Coleman HM, Brierley I, Stevenson PG. 2003. An internal ribosome entry site directs translation of the murine gammaherpesvirus 68 MK3 open reading frame. *J Virol 77*:13093-13105.

Cornelis S, Bruynooghe Y, Denecker G, Van Huffel S, Tinton S, Beyaert R. 2000. Identification and characterization of a novel cell cycle-regulated internal ribosome entry site. *Mol Cell 5*:597-605.

Czibener C, Alvarez D, Scodeller E, Gamarnik AV. 2005. Characterization of internal ribosomal entry sites of Triatoma virus. *J Gen Virol 86*:2275-2280.

Dahme T, Wood J, Livingston DM, Gaubatz S. 2002. Two different E2F6 proteins generated by alternative splicing and internal translation initiation. *Eur J Biochem 269*:5030-5036.

De Pietri Tonelli D, Mihailovich M, Schnurbus R, Pesole G, Grohovaz F, Zacchetti D. 2003. Translational control of Scamper expression via a cell-specific internal ribosome entry site. *Nucleic Acids Res 31*:2508-2513.

Deffaud C, Darlix JL. 2000. Rous sarcoma virus translation revisited: characterization of an internal ribosome entry segment in the 5' leader of the genomic RNA. *J Virol 74*:11581-11588.

Dinkova TD, Zepeda H, Martinez-Salas E, Martinez LM, Nieto-Sotelo J, de Jimenez ES. 2005. Cap-independent translation of maize Hsp101. *Plant J 41*:722-731.

Dobson T, Minic A, Nielsen K, Amiott E, Krushel L. 2005. Internal initiation of translation of the TrkB mRNA is mediated by multiple regions within the 5' leader. *Nucleic Acids Res 33*:2929-2941.

Domier LL, McCoppin NK, D'Arcy CJ. 2000. Sequence requirements for translation initiation of Rhopalosiphum padi virus ORF2. *Virology 268*:264-271.

Donnelly ML, Hughes LE, Luke G, Mendoza H, ten Dam E, Gani D, Ryan MD. 2001a. The 'cleavage' activities of foot-and-mouth disease virus 2A site-directed mutants and naturally occurring '2A-like' sequences. *J Gen Virol 82*:1027-1041.

Donnelly ML, Luke G, Mehrotra A, Li X, Hughes LE, Gani D, Ryan MD. 2001b. Analysis of the aphthovirus 2A/2B polyprotein 'cleavage' mechanism indicates not a proteolytic reaction, but a novel translational effect: a putative ribosomal 'skip'. *J Gen Virol 82*:1013-1025.

Dorner AJ, Semler BL, Jackson RJ, Hanecak R, Duprey E, Wimmer E. 1984. In vitro translation of poliovirus RNA: utilization of internal initiation sites in reticulocyte lysate. *J Virol 50*:507-514.

Dreher TW, Miller WA. 2006. Translational control in positive strand RNA plant viruses. *Virology 344*:185-197.

Dyer JR, Michel S, Lee W, Castellucci VF, Wayne NL, Sossin WS. 2003. An activity-dependent switch to cap-independent translation triggered by eIF4E dephosphorylation. *Nat Neurosci 6*:219-220.

Fabian MR, White KA. 2004. 5'-3' RNA-RNA interaction facilitates cap- and poly(A) tail-independent translation of tomato bushy stunt virus mrna: a potential common mechanism for tombusviridae. *J Biol Chem 279*:28862-28872.

Fabian MR, White KA. 2006. Analysis of a 3'-translation enhancer in a tombusvirus: a dynamic model for RNA-RNA interactions of mRNA termini. *Rna 12*:1304-1314.

Ferguson RJ, De Morais SM, Benhamou S, Bouchardy C, Blaisdell J, Ibeanu G, Wilkinson GR, Sarich TC, Wright JM, Dayer P, Goldstein JA. 1998. A new genetic defect in human CYP2C19: mutation of the initiation codon is responsible for poor metabolism of S-mephenytoin. *J Pharmacol Exp Ther 284*:356-361.

Fernandez J, Yaman I, Mishra R, Merrick WC, Snider MD, Lamers WH, Hatzoglou M. 2001. Internal ribosome entry site-mediated translation of a mammalian mRNA is regulated by amino acid availability. *J Biol Chem 276*:12285-12291.

Fronzes R, Chaignepain S, Bathany K, Giraud MF, Arselin G, Schmitter JM, Dautant A, Velours J, Brethes D. 2003. Topological and functional study of subunit h of the F1Fo ATP synthase complex in yeast Saccharomyces cerevisiae. *Biochemistry 42*:12038-12049.

Gallie DR. 2002. The 5'-leader of tobacco mosaic virus promotes translation through enhanced recruitment of eIF4F. *Nucleic Acids Res 30*:3401-3411.

Gan W, Rhoads RE. 1996. Internal initiation of translation directed by the 5'-untranslated region of the mRNA for eIF4G, a factor involved in the picornavirus-induced switch from cap-dependent to internal initiation. *J Biol Chem 271*:623-626.

Garlapati S, Wang CC. 2004. Identification of a novel internal ribosome entry site in giardiavirus that extends to both sides of the initiation codon. *J Biol Chem 279*:3389-3397.

Gazo BM, Murphy P, Gatchel JR, Browning KS. 2004. A novel interaction of Cap-binding protein complexes eukaryotic initiation factor (eIF) 4F and eIF(iso)4F with a region in the 3'-untranslated region of satellite tobacco necrosis virus. *J Biol Chem 279*:13584-13592.

Gharbi J, el Hiar R, Ben M'hadheb M, Jaidane H, Bouslama L, N'Saibia S, Aouni M. 2006. Nucleotide sequences of IRES domains IV and V of natural ECHO virus type 11 isolates with different replicative capacity phenotypes. *Virus Genes 32*:269-276.

Giersing B, Miura K, Shimp R, Wang J, Zhou H, Orcutt A, Stowers A, Saul A, Miller LH, Long C, Singh S. 2005. posttranslational modification of recombinant Plasmodium falciparum apical membrane antigen 1: impact on functional immune responses to a malaria vaccine candidate. *Infect Immun 73*:3963-3970.

Grace K, Gartland M, Karayiannis P, McGarvey MJ, Clarke B. 1999. The 5' untranslated region of GB virus B shows functional similarity to the internal ribosome entry site of hepatitis C virus. *J Gen Virol 80 (Pt 9)*:2337-2341.

Griffiths A, Coen DM. 2005. An unusual internal ribosome entry site in the herpes simplex virus thymidine kinase gene. *Proc Natl Acad Sci U S A 102*:9667-9672.

Grobe K, Esko JD. 2002. Regulated translation of heparan sulfate N-acetylglucosamine N-deacetylase/n-sulfotransferase isozymes by structured 5'-untranslated regions and internal ribosome entry sites. *J Biol Chem 277*:30699-30706.

Guo L, Allen EM, Miller WA. 2001. Base-pairing between untranslated regions facilitates translation of uncapped, nonpolyadenylated viral RNA. *Mol Cell 7*:1103-1109.

Han B, Dong Z, Zhang JT. 2003. Tight control of platelet-derived growth factor B/c-sis expression by interplay between the 5'-untranslated region sequence and the major upstream promoter. *J Biol Chem 278*:46983-46993.

Han B, Zhang JT. 2002. Regulation of gene expression by internal ribosome entry sites or cryptic promoters: the eIF4G story. *Mol Cell Biol 22*:7372-7384.

Han F, Zhang X. 2006. Internal initiation of mRNA translation in insect cell mediated by an internal ribosome entry site (IRES) from shrimp white spot syndrome virus (WSSV). *Biochem Biophys Res Commun 344*:893-899.

Hatakeyama Y, Shibuya N, Nishiyama T, Nakashima N. 2004. Structural variant of the intergenic internal ribosome entry site elements in dicistroviruses and computational search for their counterparts. *Rna 10*:779-786.

Hecht K, Bailey JE, Minas W. 2002. Polycistronic gene expression in yeast versus cryptic promoter elements. *FEMS Yeast Res 2*:215-224.

Henis-Korenblit S, Strumpf NL, Goldstaub D, Kimchi A. 2000. A novel form of DAP5 protein accumulates in apoptotic cells as a result of caspase cleavage and internal ribosome entry site-mediated translation. *Mol Cell Biol 20*:496-506.

Heppner Goss K, Trzepacz C, Tuohy TM, Groden J. 2002. Attenuated APC alleles produce functional protein from internal translation initiation. *Proc Natl Acad Sci U S A 99*:8161-8166.

Herbreteau CH, Weill L, Decimo D, Prevot D, Darlix JL, Sargueil B, Ohlmann T. 2005. HIV-2 genomic RNA contains a novel type of IRES located downstream of its initiation codon. *Nat Struct Mol Biol 12*:1001-1007.

Hernandez-Munoz I, Benet M, Calero M, Jimenez M, Diaz R, Pellicer A. 2003. rgr oncogene: activation by elimination of translational controls and mislocalization. *Cancer Res 63*:4188-4195.

Hernandez G, Vazquez-Pianzola P, Sierra JM, Rivera-Pomar R. 2004. Internal ribosome entry site drives cap-independent translation of reaper and heat shock protein 70 mRNAs in Drosophila embryos. *Rna 10*:1783-1797.

Hershey JWB, Merrick WC. 2000. Pathway and mechanism of initiation of protein synthesis. In: Sonenberg N, Hershey JWB, Mathews MB, eds. *Translational control of gene expression*. Cold Spring Harbor, NY.: Cold Spring Harbor Laboratory Press. pp 33-88.

Hinton TM, Crabb BS. 2001. The novel picornavirus Equine rhinitis B virus contains a strong type II internal ribosomal entry site which functions similarly to that of Encephalomyocarditis virus. *J Gen Virol 82*:2257-2269.

Hinton TM, Li F, Crabb BS. 2000. Internal ribosomal entry site-mediated translation initiation in equine rhinitis A virus: similarities to and differences from that of foot-and-mouth disease virus. *J Virol 74*:11708-11716.

Holcik M, Sonenberg N. 2005. Translational control in stress and apoptosis. *Nat Rev Mol Cell Biol 6*:318-327.

Hudder A, Werner R. 2000. Analysis of a Charcot-Marie-Tooth disease mutation reveals an essential internal ribosome entry site element in the connexin-32 gene. *J Biol Chem 275*:34586-34591.

Huez I, Creancier L, Audigier S, Gensac MC, Prats AC, Prats H. 1998. Two independent internal ribosome entry sites are involved in translation initiation of vascular endothelial growth factor mRNA. *Mol Cell Biol 18*:6178-6190.

Hunnicutt LE, Hunter WB, Cave RD, Powell CA, Mozoruk JJ. 2006. Genome sequence and molecular characterization of Homalodisca coagulata virus-1, a novel virus discovered in the glassy-winged sharpshooter (Hemiptera: Cicadellidae). *Virology 350*:67-78.

Chappell SA, Edelman GM, Mauro VP. 2000. A 9-nt segment of a cellular mRNA can function as an internal ribosome entry site (IRES) and when present in linked multiple copies greatly enhances IRES activity. *Proc Natl Acad Sci U S A 97*:1536-1541.

Chappell SA, Owens GC, Mauro VP. 2001. A 5' Leader of Rbm3, a Cold Stress-induced mRNA, Mediates Internal Initiation of Translation with Increased Efficiency under Conditions of Mild Hypothermia. *J Biol Chem 276*:36917-36922.

Chard LS, Bordeleau ME, Pelletier J, Tanaka J, Belsham GJ. 2006. Hepatitis C virus-related internal ribosome entry sites are found in multiple genera of the family Picornaviridae. *J Gen Virol 87*:927-936.

Chartrand P, Meng XH, Huttelmaier S, Donato D, Singer RH. 2002. Asymmetric sorting of ash1p in yeast results from inhibition of translation by localization elements in the mRNA. *Mol Cell 10*:1319-1330.

Cherry S, Doukas T, Armknecht S, Whelan S, Wang H, Sarnow P, Perrimon N. 2005. Genome-wide RNAi screen reveals a specific sensitivity of IRES-containing RNA viruses to host translation inhibition. *Genes Dev 19*:445-452.

Chiang PW, Carpenter LE, Hagerman PJ. 2001. The 5'-untranslated region of the FMR1 message facilitates translation by internal ribosome entry. *J Biol Chem 276*:37916-37921.

Iizuka N, Najita L, Franzusoff A, Sarnow P. 1994. Cap-dependent and cap-independent translation by internal initiation of mRNAs in cell extracts prepared from Saccharomyces cerevisiae. *Mol Cell Biol 14*:7322-7330.

Isaksson A, Berggren M, Ricksten A. 2003. Epstein-Barr virus U leader exon contains an internal ribosome entry site. *Oncogene 22*:572-581.

Ivanov PA, Karpova OV, Skulachev MV, Tomashevskaya OL, Rodionova NP, Dorokhov Yu L, Atabekov JG. 1997. A tobamovirus genome that contains an internal ribosome entry site functional in vitro. *Virology 232*:32-43.

Jaag HM, Kawchuk L, Rohde W, Fischer R, Emans N, Prufer D. 2003. An unusual internal ribosomal entry site of inverted symmetry directs expression of a potato leafroll polerovirus replication-associated protein. *Proc Natl Acad Sci U S A 100*:8939-8944.

Jackson RJ. 2000. Comparative view of initiation site selection mechanisms. In: Sonenberg N, Hershey JWB, Mathews MB, eds. *Translational control of gene expression.* Cold Spring Harbor, NY.: Cold Spring Harbor Laboratory Press. pp 127-183.

Jang GM, Leong LE, Hoang LT, Wang PH, Gutman GA, Semler BL. 2004. Structurally distinct elements mediate internal ribosome entry within the 5'-noncoding region of a voltage-gated potassium channel mRNA. *J Biol Chem 279*:47419-47430.

Jang SK, Krausslich HG, Nicklin MJ, Duke GM, Palmenberg AC, Wimmer E. 1988. A segment of the 5' nontranslated region of encephalomyocarditis virus RNA directs internal entry of ribosomes during in vitro translation. *J Virol 62*:2636-2643.

Jimenez J, Jang GM, Semler BL, Waterman ML. 2005. An internal ribosome entry site mediates translation of lymphoid enhancer factor-1. *Rna 11*:1385-1399.

Johannes G, Carter MS, Eisen MB, Brown PO, Sarnow P. 1999. Identification of eukaryotic mRNAs that are translated at reduced cap binding complex eIF4F concentrations using a cDNA microarray. *Proc Natl Acad Sci U S A 96*:13118-13123.

Jopling CL, Spriggs KA, Mitchell SA, Stoneley M, Willis AE. 2004. L-Myc protein synthesis is initiated by internal ribosome entry. *Rna 10*:287-298.

Jopling CL, Willis AE. 2001. N-myc translation is initiated via an internal ribosome entry segment that displays enhanced activity in neuronal cells. *Oncogene 20*:2664-2670.

Kaku Y, Chard LS, Inoue T, Belsham GJ. 2002. Unique characteristics of a picornavirus internal ribosome entry site from the porcine teschovirus-1 talfan. *J Virol 76*:11721-11728.

Kean KM. 2003. The role of mRNA 5'-noncoding and 3'-end sequences on 40S ribosomal subunit recruitment, and how RNA viruses successfully compete with cellular mRNAs to ensure their own protein synthesis. *Biol Cell 95*:129-139.

Kim JG, Armstrong RC, Berndt JA, Kim NW, Hudson LD. 1998. A secreted DNA-binding protein that is translated through an internal ribosome entry site (IRES) and distributed in a discrete pattern in the central nervous system. *Mol Cell Neurosci 12*:119-140.

Kim JH, Paek KY, Ha SH, Cho S, Choi K, Kim CS, Ryu SH, Jang SK. 2004. A cellular RNA-binding protein enhances internal ribosomal entry site-dependent translation through an interaction downstream of the hepatitis C virus polyprotein initiation codon. *Mol Cell Biol 24*:7878-7890.

Kneller EL, Rakotondrafara AM, Miller WA. 2006. Cap-independent translation of plant viral RNAs. *Virus Res 119*:63-75.

Koh DC, Wong SM, Liu DX. 2003. Synergism of the 3'-untranslated region and an internal ribosome entry site differentially enhances the translation of a plant virus coat protein. *J Biol Chem 278*:20565-20573.

Komar AA, Hatzoglou M. 2005. Internal ribosome entry sites in cellular mRNAs: mystery of their existence. *J Biol Chem 280*:23425-23428.

Komar AA, Lesnik T, Cullin C, Merrick WC, Trachsel H, Altmann M. 2003. Internal initiation drives the synthesis of Ure2 protein lacking the prion domain and affects [URE3] propagation in yeast cells. *Embo J 22*:1199-1209.

Kozak M. 1989. The scanning model for translation: an update. *J Cell Biol 108*:229-241.

Kozak M. 2001. New ways of initiating translation in eukaryotes? *Mol Cell Biol 21*:1899-1907.

Kozak M. 2005. A second look at cellular mRNA sequences said to function as internal ribosome entry sites. *Nucleic Acids Res 33*:6593-6602.

Kuhn R, Luz N, Beck E. 1990. Functional analysis of the internal translation initiation site of foot-and-mouth disease virus. *J Virol 64*:4625-4631.

Lahlou H, Fanjul M, Pradayrol L, Susini C, Pyronnet S. 2005. Restoration of functional gap junctions through internal ribosome entry site-dependent synthesis of endogenous connexins in density-inhibited cancer cells. *Mol Cell Biol 25*:4034-4045.

Lang KJ, Kappel A, Goodall GJ. 2002. Hypoxia-inducible factor-1alpha mRNA contains an internal ribosome entry site that allows efficient translation during normoxia and hypoxia. *Mol Biol Cell 13*:1792-1801.

Lauring AS, Overbaugh J. 2000. Evidence that an IRES within the Notch2 coding region can direct expression of a nuclear form of the protein. *Mol Cell 6*:939-945.

Le Quesne JP, Stoneley M, Fraser GA, Willis AE. 2001. Derivation of a structural model for the c-myc IRES. *J Mol Biol 310*:111-126.

Le SY, Maizel JV, Jr. 1997. A common RNA structural motif involved in the internal initiation of translation of cellular mRNAs. *Nucleic Acids Res 25*:362-369.

Lee JC, Wu TY, Huang CF, Yang FM, Shih SR, Hsu JT. 2005. High-efficiency protein expression mediated by enterovirus 71 internal ribosome entry site. *Biotechnol Bioeng 90*:656-662.

Li PW, Li J, Timmerman SL, Krushel LA, Martin SL. 2006. The dicistronic RNA from the mouse LINE-1 retrotransposon contains an internal ribosome entry site upstream of each ORF: implications for retrotransposition. *Nucleic Acids Res 34*:853-864.

Liu Z, Dong Z, Han B, Yang Y, Liu Y, Zhang JT. 2005. Regulation of expression by promoters versus internal ribosome entry site in the 5'-untranslated sequence of the human cyclin-dependent kinase inhibitor p27kip1. *Nucleic Acids Res 33*:3763-3771.

Lopez-Lastra M, Gabus C, Darlix JL. 1997. Characterization of an internal ribosomal entry segment within the 5' leader of avian reticuloendotheliosis virus type A RNA and development of novel MLV-REV-based retroviral vectors. *Hum Gene Ther 8*:1855-1865.

Lopez-Lastra M, Ulrici S, Gabus C, Darlix JL. 1999. Identification of an internal ribosome entry segment in the 5' region of the mouse VL30 retrotransposon and its use in the development of retroviral vectors. *J Virol 73*:8393-8402.

Macejak DG, Sarnow P. 1991. Internal initiation of translation mediated by the 5' leader of a cellular mRNA. *Nature 353*:90-94.

Martin MM, Garcia JA, McFarland JD, Duffy AA, Gregson JP, Elton TS. 2003. Translation of the human angiotensin II type 1 receptor mRNA is mediated by a highly efficient internal ribosome entry site. *Mol Cell Endocrinol 212*:51-61.

Martineau Y, Le Bec C, Monbrun L, Allo V, Chiu IM, Danos O, Moine H, Prats H, Prats AC. 2004. Internal ribosome entry site structural motifs conserved among mammalian fibroblast growth factor 1 alternatively spliced mRNAs. *Mol Cell Biol 24*:7622-7635.

Mathews MB, Sonenberg N, Hershey JWB. 2000. Origins and principles of translational control. In: Sonenberg N, Hershey JWB, Mathews MB, eds. *Translational control of gene expression*. Cold Spring Harbor, NY.: Cold Spring Harbor Laboratory Press. pp 1-31.

Mauro VP, Edelman GM, Zhou W. 2004. Reevaluation of the conclusion that IRES-activity reported within the 5' leader of the TIF4631 gene is due to promoter activity. *Rna 10*:895-897; discussion 898.

Meignin C, Bailly JL, Arnaud F, Dastugue B, Vaury C. 2003. The 5' untranslated region and Gag product of Idefix, a long terminal repeat-retrotransposon from Drosophila melanogaster, act together to initiate a switch between translated and untranslated states of the genomic mRNA. *Mol Cell Biol 23*:8246-8254.

Merrick WC. 2004. Cap-dependent and cap-independent translation in eukaryotic systems. *Gene 332*:1-11.

Meulewaeter F, van Lipzig R, Gultyaev AP, Pleij CW, Van Damme D, Cornelissen M, van Eldik G. 2004. Conservation of RNA structures enables TNV and BYDV 5' and 3' elements to cooperate synergistically in cap-independent translation. *Nucleic Acids Res 32*:1721-1730.

Miskimins WK, Wang G, Hawkinson M, Miskimins R. 2001. Control of cyclin-dependent kinase inhibitor p27 expression by cap-independent translation. *Mol Cell Biol 21*:4960-4967.

Mitchell SA, Brown EC, Coldwell MJ, Jackson RJ, Willis AE. 2001. Protein factor requirements of the Apaf-1 internal ribosome entry segment: roles of polypyrimidine tract binding protein and upstream of N-ras. *Mol Cell Biol 21*:3364-3374.

Mizumoto H, Tatsuta M, Kaido M, Mise K, Okuno T. 2003. Cap-independent translational enhancement by the 3' untranslated region of red clover necrotic mosaic virus RNA1. *J Virol 77*:12113-12121.

Mokrejš M, Vopálenský V, Kolenatý O, Mašek T, Feketová Z, Sekyrová P, Škaloudová B, Kříž V, Pospíšek M. 2006. IRESite: the database of experimentally verified IRES structures (*www.iresite.org*). *Nucleic Acids Res 34*:D125-130.

Morrish BC, Rumsby MG. 2002. The 5' untranslated region of protein kinase Cdelta directs translation by an internal ribosome entry segment that is most active in densely growing cells and during apoptosis. *Mol Cell Biol 22*:6089-6099.

Myers LK, Myllyharju J, Nokelainen M, Brand DD, Cremer MA, Stuart JM, Bodo M, Kivirikko KI, Kang AH. 2004. Relevance of posttranslational modifications for the arthritogenicity of type II collagen. *J Immunol 172*:2970-2975.

Nanbru C, Lafon I, Audigier S, Gensac MC, Vagner S, Huez G, Prats AC. 1997. Alternative translation of the proto-oncogene c-myc by an internal ribosome entry site. *J Biol Chem 272*:32061-32066.

Negulescu D, Leong LE, Chandy KG, Semler BL, Gutman GA. 1998. Translation initiation of a cardiac voltage-gated potassium channel by internal ribosome entry. *J Biol Chem 273*:20109-20113.

Niepel M, Gallie DR. 1999. Identification and characterization of the functional elements within the tobacco etch virus 5' leader required for cap-independent translation. *J Virol 73*:9080-9088.

Oh SK, Scott MP, Sarnow P. 1992. Homeotic gene Antennapedia mRNA contains 5'-noncoding sequences that confer translational initiation by internal ribosome binding. *Genes Dev 6*:1643-1653.

Ohlmann T, Lopez-Lastra M, Darlix JL. 2000. An internal ribosome entry segment promotes translation of the simian immunodeficiency virus genomic RNA. *J Biol Chem 275*:11899-11906.

Oltean S, Banerjee R. 2005. A B12-responsive internal ribosome entry site (IRES) element in human methionine synthase. *J Biol Chem 280*:32662-32668.

Oumard A, Hennecke M, Hauser H, Nourbakhsh M. 2000. Translation of NRF mRNA is mediated by highly efficient internal ribosome entry. *Mol Cell Biol 20*:2755-2759.

Owens GC, Chappell SA, Mauro VP, Edelman GM. 2001. Identification of two short internal ribosome entry sites selected from libraries of random oligonucleotides. *Proc Natl Acad Sci U S A 98*:1471-1476.

Packham G, Brimmell M, Cleveland JL. 1997. Mammalian cells express two differently localized Bag-1 isoforms generated by alternative translation initiation. *Biochem J 328 (Pt 3)*:807-813.

Pedersen SK, Christiansen J, Hansen TO, Larsen MR, Nielsen FC. 2002. Human insulin-like growth factor II leader 2 mediates internal initiation of translation. *Biochem J 363*:37-44.

Pelletier J, Sonenberg N. 1988. Internal initiation of translation of eukaryotic mRNA directed by a sequence derived from poliovirus RNA. *Nature 334*:320-325.

Pesole G, Liuni S, Grillo G, Saccone C. 1998. UTRdb: a specialized database of 5'- and 3'-untranslated regions of eukaryotic mRNAs. *Nucleic Acids Res 26*:192-195.

Pilipenko EV, Pestova TV, Kolupaeva VG, Khitrina EV, Poperechnaya AN, Agol VI, Hellen CU. 2000. A cell cycle-dependent protein serves as a template-specific translation initiation factor. *Genes Dev 14*:2028-2045.

Pinkstaff JK, Chappell SA, Mauro VP, Edelman GM, Krushel LA. 2001. Internal initiation of translation of five dendritically localized neuronal mRNAs. *Proc Natl Acad Sci U S A 98*:2770-2775.

Poole TL, Wang C, Popp RA, Potgieter LN, Siddiqui A, Collett MS. 1995. Pestivirus translation initiation occurs by internal ribosome entry. *Virology 206*:750-754.

Pozner A, Goldenberg D, Negreanu V, Le SY, Elroy-Stein O, Levanon D, Groner Y. 2000. Transcription-coupled translation control of AML1/RUNX1 is mediated by cap- and internal ribosome entry site-dependent mechanisms. *Mol Cell Biol 20*:2297-2307.

Pyronnet S, Pradayrol L, Sonenberg N. 2000. A cell cycle-dependent internal ribosome entry site. *Mol Cell 5*:607-616.

Qin X, Sarnow P. 2004. Preferential translation of internal ribosome entry site-containing mRNAs during the mitotic cycle in mammalian cells. *J Biol Chem 279*:13721-13728.

Qu F, Morris TJ. 2000. Cap-independent translational enhancement of turnip crinkle virus genomic and subgenomic RNAs. *J Virol 74*:1085-1093.

Rabadan-Diehl C, Volpi S, Nikodemova M, Aguilera G. 2003. Translational regulation of the vasopressin v1b receptor involves an internal ribosome entry site. *Mol Endocrinol 17*:1959-1971.

Ray PS, Grover R, Das S. 2006. Two internal ribosome entry sites mediate the translation of p53 isoforms. *EMBO Rep 7*:404-410.

Rijnbrand R, van der Straaten T, van Rijn PA, Spaan WJ, Bredenbeek PJ. 1997. Internal entry of ribosomes is directed by the 5' noncoding region of classical swine fever virus and is dependent on the presence of an RNA pseudoknot upstream of the initiation codon. *J Virol 71*:451-457.

Ronfort C, De Breyne S, Sandrin V, Darlix JL, Ohlmann T. 2004. Characterization of two distinct RNA domains that regulate translation of the Drosophila gypsy retroelement. *Rna 10*:504-515.

Rubtsova MP, Sizova DV, Dmitriev SE, Ivanov DS, Prassolov VS, Shatsky IN. 2003. Distinctive properties of the 5'-untranslated region of human hsp70 mRNA. *J Biol Chem 278*:22350-22356.

Sakoda Y, Ross-Smith N, Inoue T, Belsham GJ. 2001. An attenuating mutation in the 2A protease of swine vesicular disease virus, a picornavirus, regulates cap- and internal ribosome entry site-dependent protein synthesis. *J Virol 75*:10643-10650.

Sasaki J, Nakashima N. 1999. Translation initiation at the CUU codon is mediated by the internal ribosome entry site of an insect picorna-like virus in vitro. *J Virol 73*:1219-1226.

Sehgal A, Briggs J, Rinehart-Kim J, Basso J, Bos TJ. 2000. The chicken c-Jun 5' untranslated region directs translation by internal initiation. *Oncogene 19*:2836-2845.

Seino A, Yanagida Y, Aizawa M, Kobatake E. 2005. Translational control by internal ribosome entry site in Saccharomyces cerevisiae. *Biochim Biophys Acta 1681*:166-174.

Shen R, Miller WA. 2004. The 3' untranslated region of tobacco necrosis virus RNA contains a barley yellow dwarf virus-like cap-independent translation element. *J Virol 78*:4655-4664.

Sherrill KW, Byrd MP, Van Eden ME, Lloyd RE. 2004. BCL-2 translation is mediated via internal ribosome entry during cell stress. *J Biol Chem 279*:29066-29074.

Shi Y, Sharma A, Wu H, Lichtenstein A, Gera J. 2005. Cyclin D1 and c-myc internal ribosome entry site (IRES)-dependent translation is regulated by AKT activity and enhanced by rapamycin through a p38 MAPK- and ERK-dependent pathway. *J Biol Chem 280*:10964-10973.

Shiroki K, Ohsawa C, Sugi N, Wakiyama M, Miura K, Watanabe M, Suzuki Y, Sugano S. 2002. Internal ribosome entry site-mediated translation of Smad5 in vivo: requirement for a nuclear event. *Nucleic Acids Res 30*:2851-2861.

Schaad MC, Anderberg RJ, Carrington JC. 2000. Strain-specific interaction of the tobacco etch virus NIa protein with the translation initiation factor eIF4E in the yeast two-hybrid system. *Virology 273*:300-306.

Scheu S, Stetson DB, Reinhardt RL, Leber JH, Mohrs M, Locksley RM. 2006. Activation of the integrated stress response during T helper cell differentiation. *Nat Immunol 7*:644-651.

Schiavi A, Hudder A, Werner R. 1999. Connexin43 mRNA contains a functional internal ribosome entry site. *FEBS Lett 464*:118-122.

Schneider R, Agol VI, Andino R, Bayard F, Cavener DR, Chappell SA, Chen JJ, Darlix JL, Dasgupta A, Donze O, Duncan R, Elroy-Stein O, Farabaugh PJ, Filipowicz W, Gale M, Jr., Gehrke L, Goldman E, Groner Y, Harford JB, Hatzglou M, He B, Hellen CU, Hentze MW, Hershey J, Hershey P, Hohn T, Holcik M, Hunter CP, Igarashi K, Jackson R, Jagus R, Jefferson LS, Joshi B, Kaempfer R, Katze M, Kaufman RJ, Kiledjian M, Kimball SR, Kimchi A, Kirkegaard K, Koromilas AE, Krug RM, Kruys V, Lamphear BJ, Lemon S, Lloyd RE, Maquat LE, Martinez-Salas E, Mathews MB, Mauro VP, Miyamoto S, Mohr I, Morris DR, Moss EG, Nakashima N, Palmenberg A, Parkin NT, Pe'ery T, Pelletier J, Peltz S, Pestova TV, Pilipenko EV, Prats AC, Racaniello V, Read GS, Rhoads RE, Richter JD, Rivera-Pomar R, Rouault T, Sachs A, Sarnow P, Scheper GC, Schiff L, Schoenberg DR, Semler BL, Siddiqui A, Skern T, Sonenberg N, Tahara SM, Thomas AA, Toulme JJ, Wilusz J, Wimmer E, Witherell G, Wormington M. 2001. New ways of initiating translation in eukaryotes. *Mol Cell Biol 21*:8238-8246.

Simons JN, Desai SM, Schultz DE, Lemon SM, Mushahwar IK. 1996. Translation initiation in GB viruses A and C: evidence for internal ribosome entry and implications for genome organization. *J Virol 70*:6126-6135.

Spriggs KA, Bushell M, Mitchell SA, Willis AE. 2005. Internal ribosome entry segment-mediated translation during apoptosis: the role of IRES-trans-acting factors. *Cell Death Differ 12*:585-591.

Stajich JE, Block D, Boulez K, Brenner SE, Chervitz SA, Dagdigian C, Fuellen G, Gilbert JG, Korf I, Lapp H, Lehvaslaiho H, Matsalla C, Mungall CJ, Osborne BI, Pocock MR, Schattner P, Senger M, Stein LD, Stupka E, Wilkinson MD, Birney E. 2002. The Bioperl toolkit: Perl modules for the life sciences. *Genome Res 12*:1611-1618.

Stoneley M, Paulin FE, Le Quesne JP, Chappell SA, Willis AE. 1998. C-Myc 5' untranslated region contains an internal ribosome entry segment. *Oncogene 16*:423-428.

Stoneley M, Willis AE. 2004. Cellular internal ribosome entry segments: structures, trans-acting factors and regulation of gene expression. *Oncogene 23*:3200-3207.

Tijms MA, van Dinten LC, Gorbalenya AE, Snijder EJ. 2001. A zinc finger-containing papain-like protease couples subgenomic mRNA synthesis to genome translation in a positive-stranded RNA virus. *Proc Natl Acad Sci U S A 98*:1889-1894.

Tsukiyama-Kohara K, Iizuka N, Kohara M, Nomoto A. 1992. Internal ribosome entry site within hepatitis C virus RNA. *J Virol 66*:1476-1483.

Vagner S, Gensac MC, Maret A, Bayard F, Amalric F, Prats H, Prats AC. 1995a. Alternative translation of human fibroblast growth factor 2 mRNA occurs by internal entry of ribosomes. *Mol Cell Biol 15*:35-44.

Vagner S, Waysbort A, Marenda M, Gensac MC, Amalric F, Prats AC. 1995b. Alternative translation initiation of the Moloney murine leukemia virus mRNA controlled by internal ribosome entry involving the p57/PTB splicing factor. *J Biol Chem 270*:20376-20383.

van Eden ME, Byrd MP, Sherrill KW, Lloyd RE. 2004a. Demonstrating internal ribosome entry sites in eukaryotic mRNAs using stringent RNA test procedures. *Rna 10*:720-730.

van Eden ME, Byrd MP, Sherrill KW, Lloyd RE. 2004b. Translation of cellular inhibitor of apoptosis protein 1 (c-IAP1) mRNA is IRES mediated and regulated during cell stress. *Rna 10*:469-481.

van Munster M, Dullemans AM, Verbeek M, van Den Heuvel JF, Clerivet A, van Der Wilk F. 2002. Sequence analysis and genomic organization of Aphid lethal paralysis virus: a new member of the family Dicistroviridae. *J Gen Virol 83*:3131-3138.

Vanderhaeghen R, De Clercq R, Karimi M, Van Montagu M, Hilson P, Van Lijsebettens M. 2006. Leader sequence of a plant ribosomal protein gene with complementarity to the 18S rRNA triggers in vitro cap-independent translation. *FEBS Lett 580*:2630-2636.

Venkatesan A, Dasgupta A. 2001. Novel fluorescence-based screen to identify small synthetic internal ribosome entry site elements. *Mol Cell Biol 21*:2826-2837.

Verge V, Vonlanthen M, Masson JM, Trachsel H, Altmann M. 2004. Localization of a promoter in the putative internal ribosome entry site of the Saccharomyces cerevisiae TIF4631 gene. *Rna 10*:277-286.

Wang S, Browning KS, Miller WA. 1997. A viral sequence in the 3'-untranslated region mimics a 5' cap in facilitating translation of uncapped mRNA. *Embo J 16*:4107-4116.

Wang S, Guo L, Allen E, Miller WA. 1999. A potential mechanism for selective control of cap-independent translation by a viral RNA sequence in cis and in trans. *Rna 5*:728-738.

Wang XQ, Rothnagel JA. 2001. Post-transcriptional regulation of the gli1 oncogene by the expression of alternative 5' untranslated regions. *J Biol Chem 276*:1311-1316.

Wang Z, Weaver M, Magnuson NS. 2005. Cryptic promoter activity in the DNA sequence corresponding to the pim-1 5'-UTR. *Nucleic Acids Res 33*:2248-2258.

Warnakulasuriyarachchi D, Cerquozzi S, Cheung HH, Holcik M. 2004. Translational induction of the inhibitor of apoptosis protein HIAP2 during endoplasmic reticulum stress attenuates cell death and is mediated via an inducible internal ribosome entry site element. *J Biol Chem 279*:17148-17157.

Warnakulasuriyarachchi D, Ungureanu NH, Holcik M. 2003. The translation of an antiapoptotic protein HIAP2 is regulated by an upstream open reading frame. *Cell Death Differ 10*:899-904.

Watada H, Mirmira RG, Leung J, German MS. 2000. Transcriptional and translational regulation of beta-cell differentiation factor Nkx6.1. *J Biol Chem 275*:34224-34230.

Wheeler DL, Church DM, Edgar R, Federhen S, Helmberg W, Madden TL, Pontius JU, Schuler GD, Schriml LM, Sequeira E, Suzek TO, Tatusova TA, Wagner L. 2004. Database resources of the National Center for Biotechnology Information: update. *Nucleic Acids Res 32*:D35-40.

Wilson JE, Powell MJ, Hoover SE, Sarnow P. 2000. Naturally occurring dicistronic cricket paralysis virus RNA is regulated by two internal ribosome entry sites. *Mol Cell Biol 20*:4990-4999.

Xiao ZS, Simpson LG, Quarles LD. 2003. IRES-dependent translational control of Cbfa1/Runx2 expression. *J Cell Biochem 88*:493-505.

Yang D, Wilson JE, Anderson DR, Bohunek L, Cordeiro C, Kandolf R, McManus BM. 1997. In vitro mutational and inhibitory analysis of the cis-acting translational elements within the 5' untranslated region of coxsackievirus B3: potential targets for antiviral action of antisense oligomers. *Virology 228*:63-73.

Ye X, Fong P, Iizuka N, Choate D, Cavener DR. 1997. Ultrabithorax and Antennapedia 5' untranslated regions promote developmentally regulated internal translation initiation. *Mol Cell Biol 17*:1714-1721.

Yoon A, Peng G, Brandenburg Y, Zollo O, Xu W, Rego E, Ruggero D. 2006. Impaired control of IRES-mediated translation in X-linked dyskeratosis congenita. *Science 312*:902-906.

Yu Y, Alwine JC. 2006. 19S late mRNAs of simian virus 40 have an internal ribosome entry site upstream of the virion structural protein 3 coding sequence. *J Virol 80*:6553-6558.

Zeenko V, Gallie DR. 2005. Cap-independent translation of tobacco etch virus is conferred by an RNA pseudoknot in the 5'-leader. *J Biol Chem 280*:26813-26824.

Zeiner GM, Sturm NR, Campbell DA. 2003. The Leishmania tarentolae spliced leader contains determinants for association with polysomes. *J Biol Chem 278*:38269-38275.

Zell R, Sidigi K, Henke A, Schmidt-Brauns J, Hoey E, Martin S, Stelzner A. 1999. Functional features of the bovine enterovirus 5'-non-translated region. *J Gen Virol 80 (Pt 9)*:2299-2309.

Zhang X, Richie C, Legerski RJ. 2002. Translation of hSNM1 is mediated by an internal ribosome entry site that upregulates expression during mitosis. *DNA Repair (Amst) 1*:379-390.

Zhou W, Edelman GM, Mauro VP. 2001. Transcript leader regions of two Saccharomyces cerevisiae mRNAs contain internal ribosome entry sites that function in living cells. *Proc Natl Acad Sci U S A 98*:1531-1536.

Zhou W, Edelman GM, Mauro VP. 2003. Isolation and identification of short nucleotide sequences that affect translation initiation in Saccharomyces cerevisiae. *Proc Natl Acad Sci U S A 100*:4457-4462.

Zuidmeer L, van Leeuwen WA, Budde IK, Cornelissen J, Bulder I, Rafalska I, Besoli NT, Akkerdaas JH, Asero R, Rivas MF, Mancebo EG, van Ree R. 2005. Lipid transfer proteins from fruit: cloning, expression and quantification. *Int Arch Allergy Immunol 137*:273-281.

In: New Messenger RNA Research Communications
Editor: Lee B. Kwang, pp. 167-180

ISBN 978-1-60021-488-2

Chapter XI

A Note on the Interpretation of *in Vitro* Ribosomal Frameshift Efficiencies

Maarten H. de Smit[8], Maarten Laurs, Cornelis W. A. Pleij and René C. L. Olsthoorn

Section Genexpress, Leiden Institute of Chemistry, Leiden University, Einsteinweg 55, P.O. Box 9502, 2300 RA Leiden, The Netherlands

Abstract

Programmed ribosomal frameshifting is a common way of achieving the production of different proteins from the same mRNA, particularly in retroviruses. The efficiency of the RNA-encoded signals that induce ribosomes to switch reading frame, is usually expressed as the frameshift efficiency, *i.e.*, the percentage of the ribosomes that, upon reaching the frameshift signal, continue in the altered reading frame. In practice, this is determined either *in vitro* or *in vivo* as the amount of shifted product, divided by the sum of the shifted and non-shifted products. In the course of a study on the properties of the frameshift signal from *Simian retrovirus 1* RNA, the frameshift efficiency thus determined was found to change dramatically with time in an *in vitro* time-course experiment. Here we show that these variations in apparent frameshift efficiency are not due to varying behavior of the ribosomes at the frameshift signal, but to factors other than frameshifting, affecting the relative amounts of the two protein products. A kinetic model is presented, incorporating besides a constant probability of frameshifting also the kinetics of translation elongation over the two reading frames and the stability of the two protein products. Using our experimental data, we demonstrate that fitting the results from a time-course experiment to this model allows one to compensate for these two factors and thus arrive at more reliable values for the frameshift efficiency without the need for additional, dedicated experiments.

[8] Corresponding author. Present address: Section Biological Chemistry, *ibid.*Tel. +31 71 5274412; Fax +31 71 5274357; e-mail m.smit@chem.leidenuniv.nl.

Introduction

Programmed ribosomal frameshifting, where translating ribosomes are locally induced to move over the mRNA template by a number of nucleotides different from +3 to produce an altered peptide, was originally proposed in 1982 as a potential mechanism explaining the coupled expression of two genes from bacteriophage MS2 [Kastelein *et al.*, 1982]. Although this explanation was later proven incorrect for this particular system, the mechanism soon reappeared as one of the leading principles behind gene expression from the minimized genomes of plus-strand RNA- and retroviruses [Jacks and Varmus, 1985; for reviews, see Farabaugh, 1996a, b; Giedroc *et al.*, 2000]. Moreover, ribosomal frameshifting turned up in cellular mRNAs as well [reviewed by Namy *et al.*, 2004] and it was recently proposed to perhaps play a major role in producing the aberrant peptides linked to various neurodegenerative diseases [Wills and Atkins, 2006].

In the past 25 years, several of the RNA-encoded signals inducing high-efficiency ribosomal frameshifting have been analyzed in detail, presently culminating in the cryo-electronmicroscopic imaging of ribosomes stalled while shifting frame [Namy *et al.*, 2006]. But although the latter study has provided a general, mechanistic explanation of how the ubiquitous RNA pseudoknots at programmed frameshift sites may manipulate the ribosomes into slipping, many questions on the structural and dynamic details remain unanswered.

To compare the effectiveness of different wild-type or mutant frameshift signals, the concept of "frameshift efficiency" is commonly used. This is the percentage of the ribosomes that, upon reaching the frameshift signal, continue in an altered reading frame. Because events at the molecular level are not easily visualized, the frameshift efficiency is usually determined by measuring the relative amounts of the non-shifted and shifted protein products, either *in vitro* after a fixed incubation time, or *in vivo* at steady state or following induction. It has, however, been notoriously difficult to compare frameshift efficiencies between different laboratories, even when the same frameshift signals were used. In part, this may be due to differences in procedures, cell extracts, salt conditions, *et cetera* affecting the frameshifting process (see Conclusion). However, in this chapter we show that the value of the frameshift efficiency determined *in vitro* in this manner is also strongly affected by the dynamics of processes other than the events at the frameshift site, and thereby depends on the incubation time. Using a derivative of the frameshift signal from *Simian retrovirus 1* (SRV-1) [ten Dam *et al.*, 1994] as a model system, we have analyzed how the apparent frameshift efficiency *in vitro* is affected by two of these factors, namely the synthesis times of the two protein products and their physical stabilities. We present a kinetic model that allows compensation for these factors, through curve fitting of time-course protein-synthesis data to the mathematical equation representing this model. This approach obviates the need for separate, dedicated experiments for quantifying these factors and allows more reliable quantification of frameshifting.

Materials and Methods

Plasmid

The plasmid used (pSF226) is a derivative of pSF2 [ten Dam et al., 1994]. It carries the gene encoding the PB2 polymerase from Influenza A virus under control of the SP6 promoter. Part of the coding region was replaced by the sequence shown in Figure 1 such that 0-frame translation terminates at the indicated position after codon 164. Upon shifting to the -1 frame at the indicated slip site, the ribosomes reach the natural PB2 termination codon after having synthesized a protein of 537 amino acids.

In Vitro Translation

Samples from SP6 transcription reactions containing about 0.3 pmol of transcript were translated in a Rabbit Reticulocyte Lysate System (Promega, Madison WI, USA) according to the manufacturer's instructions, at 28°C, in a final volume of 15 μl. Proteins were labeled with 1 mCi of [^{35}S]-methionine (1175 Ci/mmol; MP Biomedicals Europe, Illkirch, France). The 0-frame and -1-frame products contain 10 and 28 methionine residues, respectively. No chase with cold methionine was performed, because this would disturb the ratio of radioactive methionines incorporated in the two proteins. To determine the physical half-lives of the proteins, hygromycin B was added after 20 minutes of protein synthesis, to a final concentration of 100 μg/ml. The protein products were separated by 12% SDS-PAGE and the incorporated radioactivity was quantified using a phosphoimager (Molecular Imager FX, Bio-Rad, Hercules CA, USA).

Results

Setting up and Validating a Detection System for Frameshifting

In setting up the necessary techniques for the analysis of mutants of the frameshift signal from SRV-1 RNA (to be reported elsewhere), we analyzed a number of variables potentially affecting frameshifting. For this purpose, a reporter plasmid was constructed in which various frameshift signals can be inserted such that the non-shifted (0-frame) polypeptide is 164 amino acids long and the shifted (-1-frame) one 537 amino acids. The difference in size allows for easy separation of the products by SDS-PAGE. SP6 RNA polymerase is used to synthesize mRNA *in vitro*. The variant used in this chapter contains a derivative of the SRV-1 frameshift signal where the slippery sequence and the pseudoknot have been replaced with the structure shown in figure 1. Frameshifting was examined by translation of SP6-transcripts in a reticulocyte lysate-based cell-free system. [^{35}S]-Methionine labeled protein products were separated by SDS-PAGE and quantified by phosphoimaging.

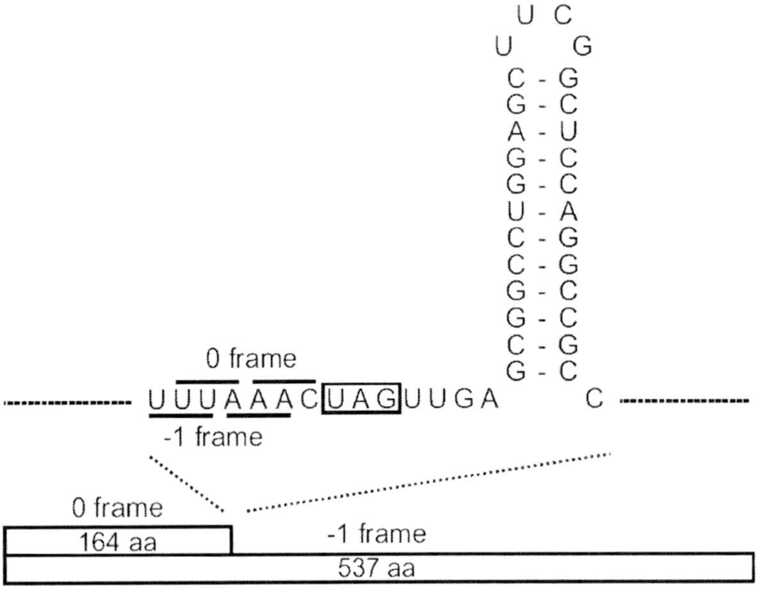

Figure 1. Open reading frames and frameshift signal. The triplet reading frames are indicated at the slip site. The termination codon of the 0 frame is boxed. The original signal from SRV-1 consists of the slippery sequence GGGAAAC and a pseudoknot structure [ten Dam *et al.*, 1994].

To account for potential differences in protein stability between the 0-frame and -1-frame products, we determined the physical half-lives of the two proteins as follows. Translation of the aforementioned mRNA was allowed to proceed for 20 minutes to produce sufficient protein for quantification. Then, hygromycin B was added to block all further synthesis and samples were taken at different time points to be analyzed as described above. figure 2 summarizes the results. Both proteins were significantly degraded during the time-course, at similar rates. The apparent frameshift efficiencies calculated from these data (amount of shifted product divided by the summed amount of shifted and non-shifted products; cross marks and gray line) decrease with time from 16% to 13%. Because no protein is being synthesized, this decrease is solely due to the slight difference in stability, causing the shifted product to disappear faster than the non-shifted product. The resulting error in the calculated frameshift efficiency is easily corrected by extrapolation to $t = 0$ and we therefore concluded that the differential stabilities of the two polypeptides would not seriously interfere with the interpretation of the frameshifting data.

Variation of the Apparent Frameshift Efficiency Over Time

Surprisingly, a very different result was obtained with the control experiment, where no hygromycin was added (figure 3). Assuming that the frameshift efficiency is an intrinsic property of the mRNA and the ribosomes, we expected here a constant value as well, representing the relative synthesis frequencies of the two peptides. In fact, a lag period was observed, followed by a dramatic increase over the first hour to a value of about 22%, significantly higher than the ~15% observed in the hygromycin experiment.

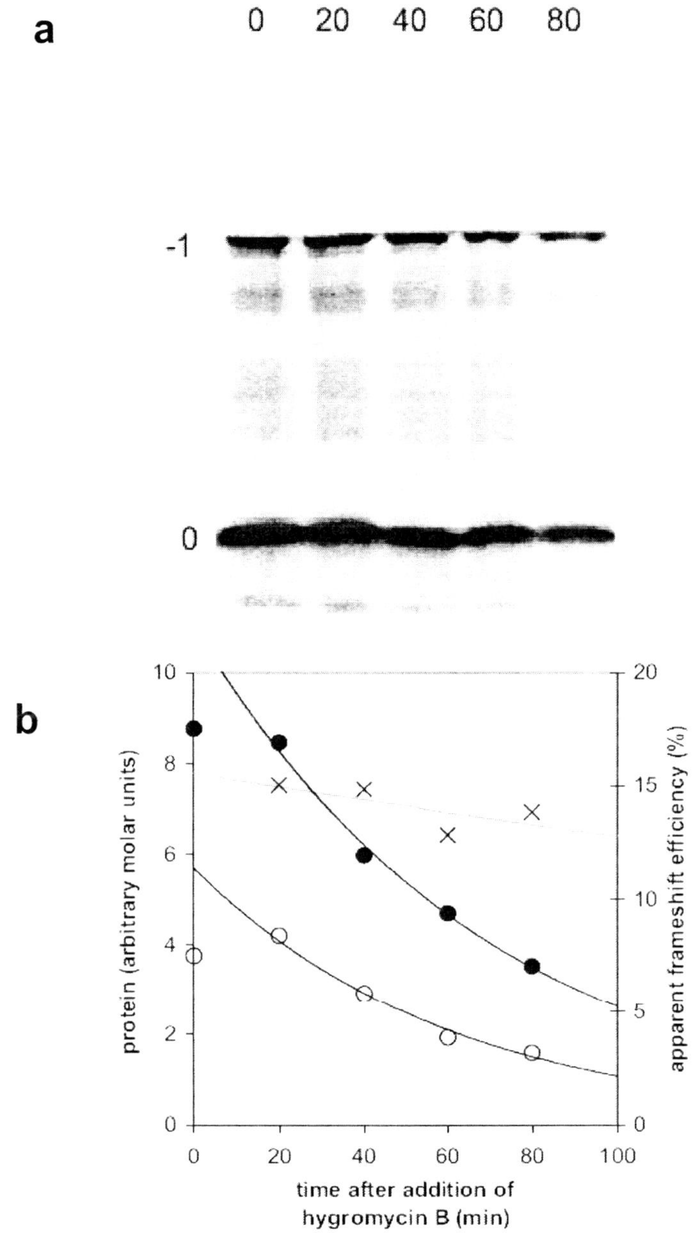

Figure 2. Protein stability. Hygromycin B was added after 20 minutes of translation. (a) Autoradiogram showing the 0-frame and -1-frame products after various incubation times as indicated. (b) Plot of the quantified data. Protein units are derived from phosphoimager counts and corrected for the background and for the number of methionine residues per molecule. Logarithmic curves denote the calculated half-lives of 49 min (0 frame, filled circles) and 42 min (-1 frame, open circles). Measurements at $t = 0$ were excluded from the calculation. Cross marks and the gray line indicate the apparent frameshift efficiencies calculated from the original measurements and the decay curves, respectively.

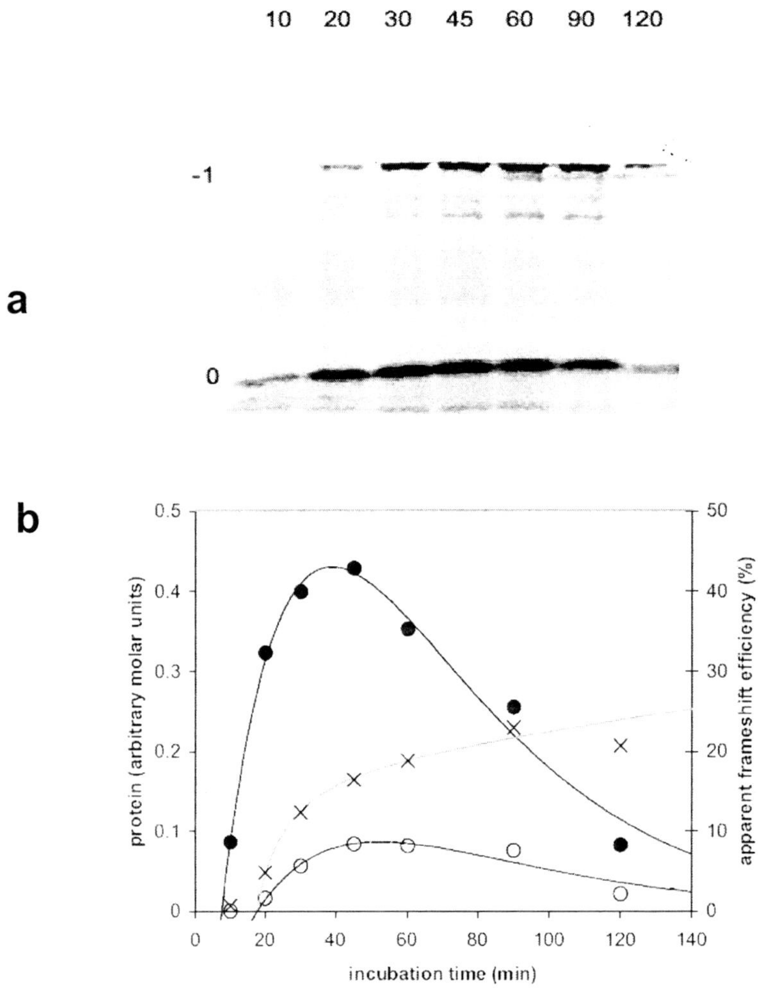

Figure 3. Variation of the apparent frameshift efficiency over time. (a) Autoradiogram showing the amounts of 0-frame and -1-frame products changing with the incubation time (min) as indicated. (b) Plot of the quantified data for 0-frame (filled circles) and -1-frame (open circles) products. Protein units are derived from phosphoimager counts and corrected for the background and for the number of methionine residues per molecule. Black curves were fitted with the kinetic model described in the text. The calculated frameshift efficiencies are indicated (cross marks and gray curve).

Assuming that we were indeed registering a change in frameshift efficiency, we initially conceived of the possibility that the rapid succession of ribosomes on the mRNA initially prevented frameshifting. The gradual decay of the activity of the extract would then lead to increasing distances between subsequent ribosomes, allowing each of them progressively more time to slip into the -1 frame. Closer inspection of the primary data from which the frameshift efficiencies were calculated, however, suggested the less exotic explanation presented next.

Variations in Apparent Frameshift Efficiency May Not Represent Changing Behavior of Ribosomes

The original autoradiogram shown in figure 3(a) appears to indicate that the 0-frame protein appears earlier than the -1-protein. Comparison of the 20' and 30' lanes indicates that this is not an illusion created by the different amounts of the two proteins, as the ratios are clearly changing over time. We therefore envisaged the possibility that a frameshifting event at the slip site was detected significantly later than a non-frameshifting event, due to the fact that in the former case a much larger protein has to be synthesized before it can be detected at the proper position in the gel (537 aa *versus* 164 aa). If this were so, the amounts of the two proteins measured at a given point in time and subsequently used to calculate the frameshift efficiency would in fact not reflect simultaneous events at the slip site.

A Kinetic Model of Processes Affecting the Apparent Frameshift Efficiency

To examine if the shapes of the curves in figure 3(b) could be understood from known factors affecting the detection of the two proteins, a kinetic model for translation of the mRNA in question was constructed, with particular focus on those factors expected to change significantly during the experiment.

(I). Decaying Initiation of Translation

Synthesis of both proteins initiates at a single translational start site. In a steady-state system, the number of proteins initiated per second would be expected to be a linear function of the mRNA concentration:

$$\frac{\mathrm{d[protein]}}{\mathrm{d}t} = k_{\mathrm{it}} \cdot [\mathrm{mRNA}] \qquad (1),$$

where k_{it} is the rate constant of translation initiation. However, it is well-known and indeed strongly suggested by Figure 3, that *in vitro* translation systems have a limited lifespan and that their activity decays with incubation time. We therefore introduced a logarithmic decay of the activity, *i.e.*, of k_{it}, with a rate constant k_{da}:

$$\frac{\mathrm{d}k_{\mathrm{it}}}{\mathrm{d}t} = -\,k_{\mathrm{da}} \cdot k_{\mathrm{it}} \qquad (2)$$

After solving these two differential equations, the total amount of protein synthesized from the translation start in question until time point t is described by:

$$[\mathrm{protein}]_{\mathrm{t}} = \frac{k_{\mathrm{it}}(0) \cdot [\mathrm{mRNA}]}{k_{\mathrm{da}}} \cdot \left(1 - e^{-k_{\mathrm{da}} \cdot t}\right) \qquad (3)$$

where $k_{it}(0)$ is the rate constant of translational initiation at the beginning of the experiment ($t = 0$). Note that the fact that the decay may at least in part be due to inactivation of the mRNA rather than of the extract is of no consequence, because k_{it} and [mRNA] are mathematically equivalent in Equation (1) and k_{da} may be considered as a composite inactivation rate constant.

(II). Delayed Detection Of Synthesis

Equation (3) calculates the synthesis level of a given protein at a given point in time as if the protein could be detected instantly, from the moment of initiation. As discussed above, however, synthesis requires a certain amount of time, depending on the ribosomal elongation rate (k_{et}, codons/min) and the number of codons to be translated (N_{cod}). This means, that detection of an initiation event is delayed by this decoding time, given by the product of the average decoding time per codon ($1/k_{et}$) and N_{cod}. In other words: protein detected at time t was initiated at time $t\text{-}N_{cod}/k_{et}$ and Equation (3) becomes:

$$[\text{protein}]_t = \frac{k_{it}(0)\cdot[\text{mRNA}]}{k_{da}}\cdot\left(1-e^{-k_{da}\cdot(t-N_{cod}/k_{et})}\right) \tag{4}$$

Proteins of different sizes (N_{cod}) will thus be detected with different delays.

(iii) Protein Degradation

The hygromycin experiment indicates that protein decay is very significant on the timescale of our experiments. As the amount of protein present at a certain time point is the resultant of synthesis and degradation (note that these are no pulse-chase experiments), this decay has to be added to the model as well. Introducing k_{pd} as the rate constant of protein degradation, Equation (1) is extended to:

$$\frac{d[\text{protein}]}{dt} = k_{it}\cdot[\text{mRNA}] - k_{pd}\cdot[\text{protein}] \tag{5}$$

As a result, equation (4) now becomes:

$$[\text{protein}]_t = \frac{k_{it}(0)\cdot[\text{mRNA}]\cdot e^{-k_{pd}\cdot(t-N_{cod}/k_{et})}}{k_{da}-k_{pd}}\cdot\left(1-e^{-(k_{da}-k_{pd})\cdot(t-N_{cod}/k_{et})}\right) \tag{6}$$

(iv). Frameshifting

Equation (6) includes all variables considered relevant to describe the detection of the synthesis of a single protein: decaying protein synthesis, delayed detection and protein degradation. One might now attempt to introduce frameshifting into the model, but this is not necessary. Assuming that there is no cross talk between shifting and non-shifting ribosomes, we can model the synthesis of the two proteins separately and independently. The different amounts of ribosomes translating the 0 frame and in the -1 frame can be simulated as different frequencies of translation initiation ($k_{it}(0)$). Initiation and frameshifting are mathematically equivalent here, because between the initiation site and the slip site, the two products do not differ in the kinetics of their synthesis.

Figure 4(a) shows an example of the output of the complete model, *i.e.*, a plot of Equation (6) for selected values of the various parameters.

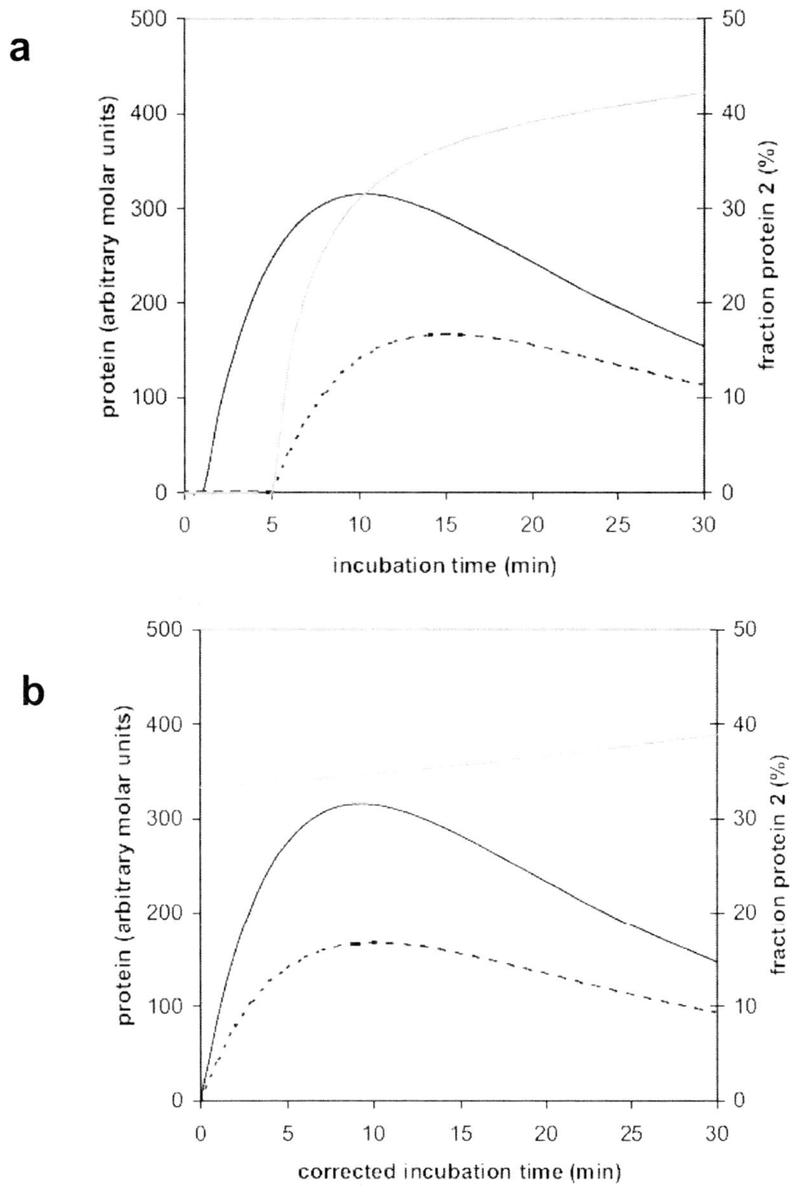

Figure 4. Output of the kinetic model for selected values of the various parameters. Protein 1 (solid black curves): $k_{it}(0)\cdot[\mathrm{mRNA}] = 100\ \mathrm{min}^{-1}$; $k_{pd} = 0.05\ \mathrm{min}^{-1}$; $N_{cod}/k_{et} = 1$ min; $k_{da} = 0.2\ \mathrm{min}^{-1}$. Protein 2 (broken curves): $k_{it}(0)\cdot[\mathrm{mRNA}] = 50\ \mathrm{min}^{-1}$; $k_{pd} = 0.04\ \mathrm{min}^{-1}$; $N_{cod}/k_{et} = 5$ min; $k_{da} = 0.2\ \mathrm{min}^{-1}$. Amounts of protein are in arbitrary molar units. Frameshift efficiencies (gray curves) were calculated from the amounts of both proteins at every point in time. (a) Real timescale. (b) Protein curves synchronized by shifting each leftward over the corresponding delay of N_{cod}/k_{et} minutes.

For illustration purposes, the values were chosen such, that their effects on the amounts of protein detected are easily distinguishable (see the legend to figure. 4). First, the different delays in the appearance of full-size proteins 1 and 2 are evident (entered as 1 and 5 minutes, respectively, to represent different-sized open reading frames). Next, the amount of protein increases rapidly in a linear fashion (the slope being defined by $k_{it}(0)\cdot[\text{mRNA}]$), but this increase soon slows down as the extract loses activity and/or the mRNA is being degraded (see above). Finally, the amount of full-length protein diminishes due to degradation, paralleling our observations in the presence of hygromycin B.

If we now assume that proteins 1 and 2 represent the 0-frame and -1-frame products in a frameshifting assay, we can see what happens if we calculate the frameshift efficiency as [protein 2]/([protein 1]+[protein 2]) (gray curve). As observed in our experiment, the apparent frameshift efficiency increases rapidly as protein 2 appears, to level off some time later. To reconstruct the true ratio between shifting and non-shifting ribosomes, we must first of all correct for the difference in decoding time between the two proteins, *i.e.*, not compare the amounts of both proteins at time point t, but at time point t-1 for protein 1 and time point t-5 for protein 2. In this way, the data are calculated back to the time of initiation (or of frameshifting, which yields the same result). The result is illustrated in figure 4(b). Because we chose a slightly longer half-life for protein 2 than for protein 1, the fraction of protein 2 now displays a moderate, monotonous increase over time (gray line), similar to Figure 2(b) where a moderate decrease was found due to the shifted product being slightly less stable than the non-shifted one. The true frameshift efficiency (33%) now equals the value of this fraction at the corrected starting point $t = 0$, before degradation has set in, and is directly and exclusively determined by the entered values for $k_{it}(0)\cdot[\text{mRNA}]$. If the protein half-lives had been chosen identical, the calculated frameshift efficiency would have remained constant at this value throughout (not shown).

The Experimental Data Can Be Fully Described by the Model

Equation 6 can now be used to examine whether our experimental data can be explained by the factors incorporated in the model and therefore, whether the true frameshift efficiency may indeed be constant over time. We used regression software to find the set of parameter values best fitting our experimental data. The result is shown in figure 5. Like in figure 4(b), the expression curves have been synchronized by shifting them leftward over the respective values found for the delay, N_{cod}/k_{et}, both for the original data (circles) and for the fitted curves (lines).

The good fit of the data points to the theoretical curves suggests that a quantitative understanding of the experimental results does not require more than the parameters incorporated in the model. However, this is only true if the values required for curve fitting are not incompatible with known data. Unfortunately, because the protein levels were measured on an arbitrary scale (based on phosphoimager counts), $k_{it}(0)$ can only be used for comparative purposes within a single experiment and cannot be verified by comparison to external data. Inspection of the curves in figure 5 reveals, however, that the calculated values of $k_{it}(0)\cdot[\text{RNA}]$ agree well with the initial slopes.

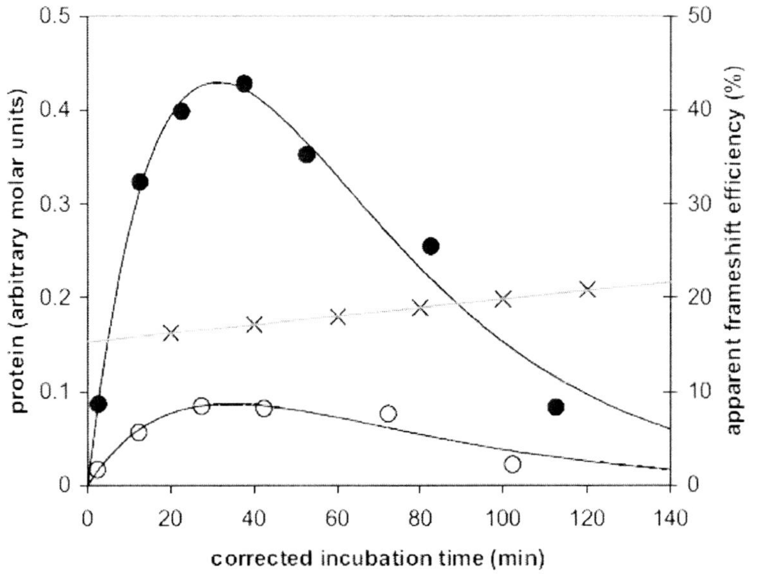

Figure 5. Fitting of the kinetic model to the experimental data (see Fig. 3). Curves were synchronized as in Figure 4(b). Fitted parameters were for the 0 frame (filled circles): $k_{it}(0)\cdot[\text{mRNA}] = 0.0374\ \text{min}^{-1}$; $k_{pd} = 0.0321\ \text{min}^{-1}$; $N_{cod}/k_{et} = 7.43$ min; $k_{da} = 0.0318\ \text{min}^{-1}$ and for the -1 frame (open circles): $k_{it}(0)\cdot[\text{mRNA}] = 0.0068\ \text{min}^{-1}$; $k_{pd} = 0.0287\ \text{min}^{-1}$; $N_{cod}/k_{et} = 17.7$ min; $k_{da} = 0.0292\ \text{min}^{-1}$. The first point of the -1-frame protein is not shown and was excluded from the curve-fitting procedure, because it lies within the lag period (figure 3).

The values found for k_{pd}, the polypeptide degradation rate constant, correspond to half-lives of 22 and 24 minutes for the 0-frame and -1-frame products, respectively. This is roughly half the values found in the hygromycin experiment. The presence of the antibiotic may have somewhat affected the physical half-life of the proteins. But given the limited number of late data points, a discrepancy by factor of two is not alarming.

The values of N_{cod}/k_{et} represent the delay between the start of the experiment and the detection of either polypeptide. If translation begins as soon as the mRNA is added to the translation mixture, k_{et} can be derived by dividing the known numbers of codons by the found parameter values. This yields values of 164/7.4 = 22 codons/min and 537/17.7 = 30 codons/min, suggesting somewhat different elongation rates over the two reading frames. Alternatively, if we consider that there may be a general lag between the addition of the mRNA and the start of methionine incorporation, the obtained values would fit a common lag of 2.8 min, followed by an overall average elongation rate of 36 codons/min. This lag must indeed be considerably shorter than the 13.9 min reported for a similar system [Lopinski et al., 2000], as the 0-frame band is already visible after 10 minutes of incubation (figure 3(a)). The calculated elongation rate compares well with reported values for reticulocyte lysates at 26°C, ranging from 28 [Lopinski et al., 2000] to 52 codons/min [Palmiter, 1973]. Pausing of a certain fraction of the ribosomes at the slip site, as observed for a frameshift signal from a yeast virus [Lopinski et al., 2000], would not dramatically alter these estimates.

Finally, the values obtained for k_{da} imply that the functional half-life of the system is 22 to 24 minutes. This value agrees very well with the data of Palmiter [1973], which fit a functional half-life of about 20 minutes for a hemin-supplemented rabbit reticulocyte lysate. The similarity between the two values obtained from the independently fitted 0-frame and -1-frame curves provides further support for the reliability of the approach. All in all, we conclude that fitting of the model to the experimental data does not require unreasonable values for the various parameters.

The frameshift efficiency can now be calculated using the obtained values for $k_{it}(0)\cdot[mRNA]$ as relative measurements of the synthesis rate of each of the two polypeptides. This results in a value of 15%. When we perform the calculation in the classical manner, using the fitted curves over their entire lengths, we find that due to the synchronization of the two proteins, the apparent frameshift efficiency no longer rises rapidly from 1% to 22% (compare the gray lines in figures 3(b) and 5). The remaining increase over time from 15% to 22% can be fully ascribed to the difference in calculated half-life between the two proteins (compare figure 4(b)). Note, however, that this is not relevant for calculation of the frameshift efficiency from the fitted values of $k_{it}(0)\cdot[mRNA]$, as shown above, because this method corresponds graphically to extrapolation of the gray line to $t = 0$.

Conclusion

Frameshift efficiencies are usually calculated by dividing the amount of the product of frameshifting by the sum of shifted and non-shifted products from the same translation start. We find, however, that the relative amounts of these two proteins at any point in time are not only determined by the probability of frameshifting. As a result, calculated frameshift efficiencies can deviate significantly from their true values.

In a model system derived from the frameshift signal of the retrovirus SRV-1, we found the calculated frameshift efficiency to vary from 1% to over 20% in a single *in vitro* time-course experiment. By constructing a mathematical model for the kinetics of *in vitro* translation and protein degradation and subsequently fitting this model to the experimental data, we arrived at a value of 15% for this specific frameshift signal under the given experimental conditions. Contrary to the initial suggestion, we do not need to postulate changes in the behavior of the ribosomes at the slip site during the course of the experiment.

Time-course experiments, combined with curve fitting, clearly provide for more accurate quantification of frameshifting than the usual single-point measurements. No reasoned choice of time point would have yielded a frameshift efficiency of 15% in our experimental setup (Fig. 3(b)). This is not to say that our findings invalidate all previous studies. The classical frameshift efficiencies are certainly useful for semiquantitative assessment of, *e.g.*, the effects of mutations within a single set of experiments. With large numbers of mutants, full-scale time-course experiments for all would become prohibitively laborious. We expect that efficiencies measured by single-point experiments may be meaningfully calibrated by performing a limited number of time-course experiments. In addition, the observed effect of delayed appearance of the synthesized proteins will be limited if the two reading frames are of similar size and the stabilities of the two polypeptides may be comparable (as in our case) if they are predominantly determined by the N-end rule of protein degradation [Varshavsky, 1992].

As an alternative to curve-fitting for the interpretation of time-course experiments, the initial rate of protein synthesis can be read from graphs like figure 3(b) as the initial slope of each curve [Lopinski *et al.*, 2000]. This would require more data points in the early, linear part of the graph, but should produce the same results, because the slope is largely insensitive both to the delayed appearance of the proteins and to their degradation. We prefer the curve-fitting approach, however, because it provides more information on the overall behavior of the system.

Our findings may provide a (partial) explanation for the fact that different laboratories often find significantly different frameshift efficiencies for the same mRNA. We have noticed that the concentrations of components of the translation mixture, like magnesium and spermidine, can profoundly affect the frameshift efficiencies calculated after a one-hour incubation (not shown). Whether this is due to a direct effect on the behavior of the ribosomes at the slip site, or on the other factors influencing the relative amounts of the two polypeptides, has not been examined. Evidently more influencing factors imply more possible causes of variation.

On a more general level, the exclusive use of derived quantities like frameshift efficiencies can clearly obscure alternative interpretations of the experimental results. We believe that this danger cannot be emphasized enough in an age where dedicated software more and more tends to present the researcher with precalculated, derived values only. Examination of the primary data remains crucial to arrive at a full understanding of the underlying processes and principles.

ACKNOWLEDGEMENTS

M.d.S. was financed by the Netherlands Organization for Scientific Research (NWO) and the Centre for Medical Systems Biology (CMSB). R.O. was financed by NWO. We thank M. Noteborn for critical reading of the manuscript.

REFERENCES

Farabaugh, P. J. (1996a). Programmed translational frameshifting. *Microbiol Rev*, *60*, 103-134.

Farabaugh, P. J. (1996b). Programmed translational frameshifting. *Annu Rev Genet*, *30*, 507-528.

Giedroc, D. P., Theimer, C. A., and Nixon, P. L. (2000). Structure, stability and function of RNA pseudoknots involved in stimulating ribosomal frameshifting. *J Mol Biol*, *298*,167-185.

Jacks, T., and Varmus, H. E. (1985). Expression of the Rous sarcoma virus pol gene by ribosomal frameshifting. *Science*, *230*, 1237–1242.

Kastelein, R. A., Remaut, E., Fiers, W., and van Duin, J. (1982). Lysis gene expression of RNA phage MS2 depends on a frameshift during translation of the overlapping coat protein gene. *Nature*, *295*, 35-41.

Lopinski, J. D., Dinman, J. D., and Bruenn, J.A. (2000). Kinetics of ribosomal pausing during programmed -1 translational frameshifting. *Mol Cell Biol*, *20*, 1095-1103.

Namy, O., Rousset, J. P., Napthine, S., and Brierley, I. (2004). Reprogrammed genetic decoding in cellular gene expression. *Mol Cell*, *13*, 157-168.

Namy, O., Moran, S. J., Stuart, D. I., Gilbert, R. J. C., and Brierley, I. (2006). A mechanical explanation of RNA pseudoknot function in programmed ribosomal frameshifting. *Nature*, *441*, 244-247.

Palmiter, R. D. (1973). Ovalbumin messenger ribonucleic acid translation. Comparable rates of polypeptide initiation and elongation on ovalbumin and globin messenger ribonucleic acid in a rabbit reticulocyte lysate. *J Biol Chem*, *248*, 2095-2106.

ten Dam, E., Brierley, I., Inglis, S., and Pleij, C. (1994). Identification and analysis of the pseudoknot-containing *gag-pro* ribosomal frameshift signal of simian retrovirus-1. *Nucleic Acids Res*, *22*, 2304-2310.

Varshavsky, A. (1992). The N-end rule. *Cell*, *69*, 725-735.

Wills, N. M., and Atkins, J. F. (2006). The potential role of ribosomal frameshifting in generating aberrant proteins implicated in neurodegenerative diseases. *RNA*, *12*, 1149-1153.

In: New Messenger RNA Research Communications
Editor: Lee B. Kwang, pp. 181-190
ISBN 978-1-60021-488-2

Chapter XII

mRNA Transfection in Monocytes and Macrophages

Tim J.L. Van De Parre, Wim Martinet and Guido R.Y. De Meyer[9]
Division of Pharmacology, University of Antwerp, Wilrijk, Belgium

Abstract

Monocytes and monocyte-derived macrophages are hard-to-transfect cells. Transfection of the human monocytic cell lines U937 and THP-1 with DNA encoding enhanced green fluorescent protein (eGFP) using standard transfection methods does not yield detectable levels of eGFP positive cells. However, nucleofector technology as a novel nonviral, electroporation-based method results in 30-50% DNA-transfected U937 or THP-1 cells, suggesting that this method is superior to others. Transfection efficiencies up to 80% without significant cell toxicity can be obtained when U937 and THP-1 cells are nucleofected with capped and polyadenylated mRNA specific for eGFP. In murine J774A.1 macrophages, introduction of DNA causes massive apoptosis without eGFP expression, whereas nucleofection with mRNA yields up to 75% eGFP positive cells without significant induction of cell death. Analogous with J774A.1 cells, primary peritoneal mouse macrophages can be easily transfected with mRNA (up to 60% eGFP positive cells) but not with DNA. Nucleofected cells start to express eGFP after an incubation period of only 2-6 hours. Despite successful transfection-driven eGFP gene expression, monocytes and macrophages fail to express for unknown reasons many other gene products upon mRNA transfection. In conclusion, transfection of monocytes and macrophages with eGFP mRNA is superior to DNA gene transfer, most likely because mRNA is less toxic and nonapoptotic as compared to plasmid DNA.

[9] Correspondence to Prof Dr Guido RY De Meyer, Division of Pharmacology, University of Antwerp, Universiteitsplein 1, B-2610 Wilrijk, Belgium, Tel. +32 3 820 27 37, Fax. +32 3 820 25 67, E-mail: Guido.Demeyer@ua.ac.be

INTRODUCTION

Human monocytes and monocyte-derived macrophages are considered to be major effectors in host defence mechanisms as well as in several pathologies such as human immunodeficiency virus infection, autoimmune diseases and chronic inflammatory disorders. Because of the expected impact on therapeutics and tissue replacement, much effort has been made to improve the transfection efficiency of these cells. But although many methods of gene transfer into mammalian cells exist, the use of genetically engineered monocytes for research purposes or therapeutic intervention has been hampered by low transfection efficiencies with nonviral vectors or by safety concerns when engineered viruses are used. Transfection with mRNA might be a valuable alternative because mRNA lacks the potential to integrate into the host genome thereby obviating safety concerns (e.g. viral recombination and insertional mutagenesis) which is an important advantage during clinical gene therapy trials [1].

In general, nonviral DNA transfection methods are inefficient, particularly in nondividing cells, as there is only very limited DNA trafficking to the nucleus where transcription occurs [2]. Several groups demonstrated the feasibility of mRNA transfection as a valid alternative for nonviral gene delivery, because this strategy avoids the need for entry into the nucleus as well as the complex issues of transcriptional regulation associated with DNA vectors [3-6]. In this chapter, we evaluate different transfection methods with DNA or mRNA in monocytes and monocyte-derived macrophages.

TRANSFECTION OF MONOCYTIC CELL LINES

Because electroporation seems to be the most promising nonviral method for efficient gene transfer into monocytes [7], this technique has been used to transfect the human monocytic cell lines U937 and THP-1 with plasmid DNA containing the enhanced green fluorescent protein (eGFP) coding sequence downstream of the human cytomegalovirus immediate early promoter. Despite uptake of plasmid DNA, the majority of transfected cells (>95%) do not express the foreign eGFP gene at a detectable level (figure 1) due to induction of cell death [8]. To explain this phenomenon, it is important to note that myeloid cells, most notably cells belonging to the monocyte/macrophage lineage, have a physiological system underlying induction of apoptosis in response to infection with many animal viruses. This feature should be regarded as a host defence mechanism to prevent propagation of microorganisms. Analogous with double-stranded RNA, which is thought to be generated during viral infection, delivery of double-stranded plasmid DNA into the cytosol of myeloid cells may lead to an antiviral state which includes the induction of cell death. An elegant solution for this problem is transfection of monocytes with mRNA on condition that the transfected mRNA contains a 5' cap structure (m^7GpppN) and 3' poly(A) tail of sufficient length. Both components are critical for mRNA stability and efficient translation [9]. The 5' cap structure activates translation by interacting with the cap-binding protein eIF4E of the translation initiation complex, whereas the poly(A) tail is needed for binding with poly (A)-binding protein 1 (PABP1). The latter protein has multiple functions including translation initiation through binding with eIF4G and mRNA stabilization [9]. Electroporation of U937 and THP-1 cells with eGFP mRNA containing a 5' m^7GpppG cap structure and poly A tail of 64 residues yields on average 60% and 45% of eGFP positive cells, respectively, suggesting that mRNA unlike DNA is non-toxic in monocytes (figure 1).

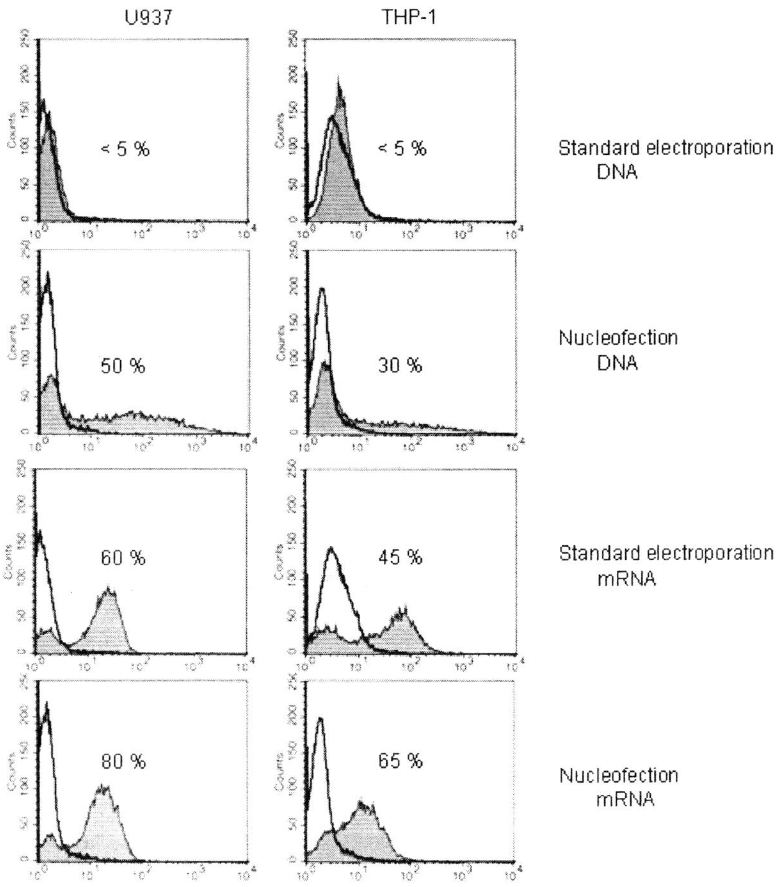

Figure 1. Flow cytometric analysis of U937 and THP-1 cells following eGFP DNA or mRNA gene transfer via standard electroporation or nucleofection. U937 and THP-1 cells were transfected with 1 µg pEGFP-N3 plasmid DNA (Clontech) or 1 µg capped and polyadenylated eGFP mRNA. Twenty-four hours after transfection, flow cytometric eGFP analysis was performed to estimate transfection efficiency of transfected cells (shaded histogram) versus controls (open histogram). The transfection efficiency is given as percentage eGFP positive cells.

In recent years, nucleofector technology is frequently used for gene delivery into hard-to-transfect cells including monocytic U937 and THP-1 cells [5]. The technology is based on standard electroporation but unlike any other transfection method DNA is transported directly into the nucleus. Typically, two different components are needed during nucleofection experiments: (i) a nucleofector device which delivers unique electrical parameters and (ii) cell-type specific solutions in which the cells are retained when the electrical program is activated. The technology is suitable for the delivery of different substrates such as DNA, mRNA or short interfering RNA without further optimization of the nucleofection conditions [10-12]. Nucleofection with plasmid DNA results in 30-50% transfected U937 or THP-1 cells (figure 1), suggesting that this method is superior to standard electroporation and other transfection techniques. Most likely, transfer of DNA directly into the nucleus inhibits the cytotoxic effects as observed with other methods [5] [8]. Transfection efficiencies increase significantly (up to 80%) when U937 and THP-1 cells are nucleofected with capped and polyadenylated mRNA (figure 1).

Transfection of Primary Monocytes

Only few studies describe successful transfection of primary monocytes in comparison to transfection of fully differentiated and matured macrophages or dendritic cells. Nucleofection with DNA of human monocytes freshly isolated from peripheral blood mononuclear cells leads to transfection efficiencies up to 60% (data from Amaxa, *http://www.amaxa.com*). By using a recovery medium, posttransfection survival of 80% can be obtained after 24 hours. However, introduction of DNA into these cells often results in limited survival over 48 hours and unresponsiveness to LPS-induced maturation limiting this technique to short-term experiments [10]

Transfection with mRNA overcomes these problems which can be explained in part by the less stringent electrical settings required for introduction of the RNA [10]. Another major advantage of mRNA transfection is that multiple mRNA species can be transfected into the target cell at the same time, leading to the translation of several different gene products within the same cell. With DNA transfection, only one or a few copies of a DNA vector will usually be incorporated, therefore limiting the expression within each cell to a few proteins [13]. However, monocytes that are first electroporated with eGFP mRNA and subsequently differentiated often have lower levels of protein expression when compared to transfection of fully matured cells [4, 14]. The low level of intact protein expression achieved by RNA transfection of primary uncultured mononuclear cells would clearly be a limitation if the intent were to modify the function of the target cell with the expressed protein. However, when the goal is to elucidate a potent immune response against the transfected antigens a specific immune response can be initiated very efficiently [13, 14].

Transfection of Macrophage Cell Lines

Transient transfection efficiency of macrophages is often low and might be due to the presence of high levels of digestive enzymes which can inactivate the transfected DNA. Moreover, transfection efficiencies are highly variable between different macrophage cell lines. RAW264.7 macrophages can be optimally transfected using a combination of lipofectamine and protamine sulfate [15]. The increased efficiency of lipofectamine over other cationic liposomes may be attributed to its greater number of positively charged groups which condense DNA more efficiently than for example the monovalent lipid lipofectin and N-[1-(2,3-Dioleoyloxy)propyl]-N,N,N-trimethylammonium methyl-sulfate (DOTAP) [15].

Protamine, which causes condensation of DNA, is added because it promotes the cellular entry and gene expression of plasmid DNA due to the presence of nuclear localization signals in its amino acid sequence [16, 17]. Previous experiments with the murine J774A.1 macrophage cell line showed that gene transfer methodologies such as electroporation, lipofection, diethylaminoethyl dextran or particle bombardment yield a transfection efficiency below 20% [18, 19]. As shown in figure 2, nucleofection of J774A.1 macrophages with eGFP plasmid DNA causes massive apoptosis without eGFP expression.

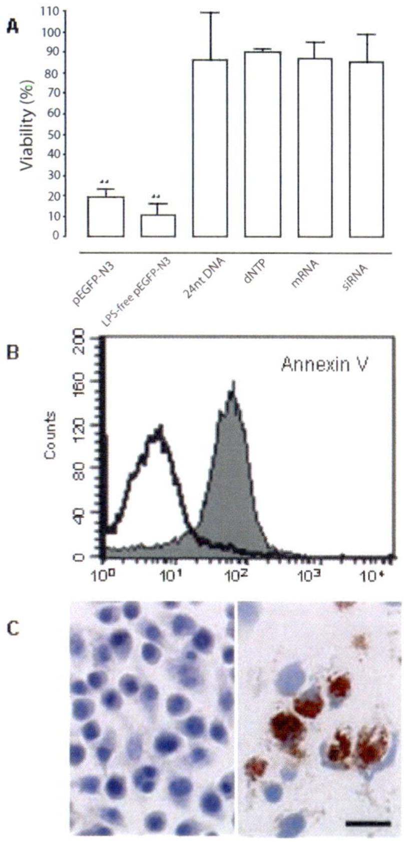

Figure 2. Characterisation of cell death 18 hours after nucleofection of J774A.1 macrophages. (A) Viability of J774A.1 cells as determined by neutral red viability assays after nucleofection of various nucleic acids. In each experiment, 10^6 macrophages were pulsed with one of the following nucleic acids (1 μg each): pEGFP-N3 plasmid (n=8), LPS-free pEGFP-N3 (n=2), 24nt double-stranded oligonucleotides (n=5), dNTP (n=2), eGFP mRNA (n=5), siRNA (n=3). ** $p<0.01$ vs. controls pulsed in the absence of nucleic acids (=100%, n=8), analysis of variance followed by the Dunnett test. Data are expressed as mean±SEM. (B) Annexin V labelling of DNA transfected cells (shaded histogram) versus controls (open histogram) as evaluated by flow cytometry. (C) Immunocytochemical detection of cleaved caspase-3 (red-brown) in control macrophages (left hand panel) and cells pulsed with 1 μg pEGFP-N3 plasmid DNA (right hand panel). Scale bar = 25 μm.

This finding suggests that macrophages are more sensitive to DNA-mediated apoptosis than monocytes. Plasmid prepared by a method designed to remove endotoxin gives an equivalent level of cell death indicating that cell death is not induced by *E. coli*-derived LPS contaminants but by DNA itself. However, addition of plasmid DNA without pulsing the cells or nucleofection with either double-stranded oligonucleotides (24 nt DNA) or dNTPs does not result in cell death (figure 2A). Besides cytotoxic effects evoked by DNA, low transfection efficiencies may result from methylation of the plasmid promoter and inhibition of transcription. Treatment of macrophages with the methylation inhibitor 5-azacytidine increases expression of plasmids driven by Rous Sarcoma Virus (RSV), SV40, and EF-1 alpha promoters [20].

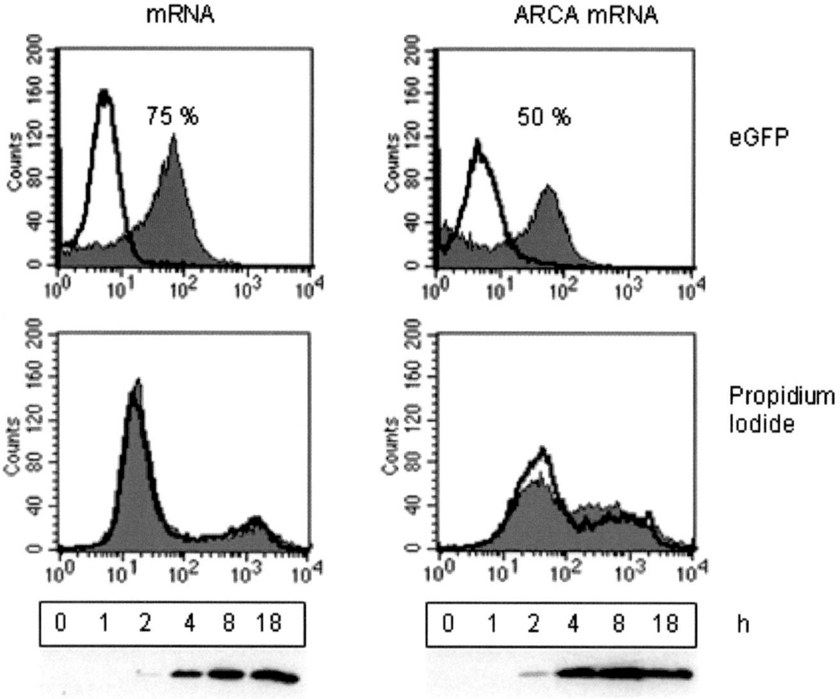

Figure 3. eGFP expression in J774A.1 macrophages after nucleofection with 2 μg eGFP-specific mRNA or ARCA mRNA (mRNA with an Anti Reverse Cap Analog). Eighteen hours after nucleofection, transfection efficiency (eGFP) and cell viability (propidium iodide) of transfected cells (shaded histogram) versus controls (open histogram) were evaluated by flow cytometry. Kinetics of eGFP expression were determined by Western blotting. Data are representative of at least four experiments. The transfection efficiency is given as percentage eGFP positive cells.

Analogous with monocytes, the difficulties observed with nucleofection of DNA in macrophages can be overcome using capped and polyadenylated mRNA. Indeed, nucleofection of J774A.1 macrophages with mRNA specific for eGFP yields up to 75% eGFP positive cells without significant induction of cell death (figure 3). Western blotting of protein lysates from transfected cells shows detectable eGFP expression as early as 2 hours after nucleofection (with maximal levels after 8 hours) (figure 3). Because initiation of translation is known to be very cap dependent, insufficient capping of in vitro synthesized eGFP mRNA may slow down protein synthesis. To improve the transfection efficiency of J774A.1 macrophages, macrophages can be nucleofected with polyadenylated mRNA containing a novel Anti Reverse Cap Analog (ARCA) [21]. In ARCA, one of the 3' OH groups (close to the 7-methylguanosine) is eliminated and substituted by an OCH_3 group. This modification allows initiation of transcription only with the remaining OH group and thus synthesis of mRNA capped exclusively in the functional, translatable orientation, theoretically resulting in higher protein yields. However, nucleofection of J774A.1 macrophages with ARCA eGFP mRNA does not result in a significant improvement of the transfection efficiency. Only a mild increase in expression of eGFP protein can be detected 2-4 hours after nucleofection (figure 3).

TRANSFECTION OF PRIMARY MACROPHAGES

Given the pivotal role of macrophages in host defence mechanisms, the use of *ex vivo*-manipulated primary macrophages is an attractive approach in the growing fields of cell-based and gene therapies. Up till now DEAE-dextran/DMSO mediated transfection of primary cultures of thioglycolate-elicited murine peritoneal macrophages has a higher efficiency with heterologous plasmid DNA than calcium-phospate or lipofection [22]. By using cochleates as a lipid-based delivery system, transgenes can be delivered to peritoneal mouse macrophages *in vitro* and *in vivo* with efficiencies up to 51 and 25%, respectively [23]. Other reported gene transfer into peritoneal macrophages is through mannose receptor-mediated endocytosis [24]. Mannose-modified chitosan appears to be a good gene carrier for macrophages. The modification of mannose to chitosan significantly enhances the transfection efficiency of the plasmid DNA/chitosan complex and reduces its cytotoxicity. Nucleofection of primary peritoneal mouse macrophages with eGFP DNA results in 20% eGFP positive cells without significant cell toxicity (figure 4). Nucleofection of murine peritoneal macrophages with eGFP mRNA increases transfection efficiencies up to 60% without cell toxicity.

NUCLEOFECTION WITH NON-EGFP-SPECIFIC MRNA

As outlined above, nucleofection with mRNA is by far the best way to transiently transfect monocytes and monocyte-derived macrophages. However, despite successful transfection-driven eGFP gene expression, monocytes and macrophages fail to express for unknown reasons many other gene products upon mRNA transfection. For example, nucleofection of macrophages with capped and polyadenylated mRNA encoding homologuous pro-apoptotic genes such as caspase-2, Death-associated Protein (DAP) kinase and BCL2/adenovirus E1B 19kD-interacting protein 3 (BNIP3) does not result in translation of the protein, as detected by western blotting (Martinet, unpublished results). Although it is conceivable that cells may not overexpress proteins that will lead to self-destruction, nucleofection with mRNA coding for other proteins such as Alzheimer disease related amyloid precursor protein (APP) or mitochondrial uncoupling protein 2 (UCP2) does not result in *de novo* expression either (Van De Parre, unpublished results). Addition of the proteasome inhibitor MG132, which prevents degradation of newly formed proteins in the cell, does not improve the transfection results. Possibly mRNA instability can explain this disturbed translation rather than degradation of the newly formed proteins after transcription.

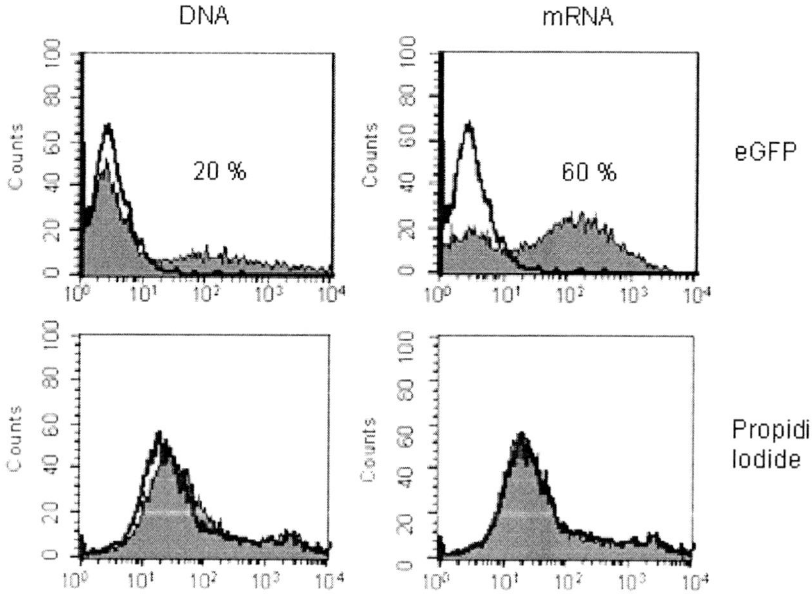

Figure 4. eGFP expression in peritoneal mouse macrophages after nucleofection with 1 μg eGFP encoding pEGFP-N3 or 2 μg eGFP-specific mRNA. Eighteen hours after nucleofection, transfection efficiency (eGFP) and cell viability (propidium iodide) of transfected cells (shaded histogram) versus controls (open histogram) were evaluated with flow cytometry.

Conclusion

In conclusion, nucleofection with mRNA is superior to transfection with DNA for the study of gene regulation as well as for the analysis of the expression and function of proteins in monocytes and macrophages, most likely because mRNA is less toxic and nonapoptotic as compared to plasmid DNA. However, despite successful transfection-driven eGFP gene expression, monocytes and macrophages fail to express for unknown reasons many other gene products upon mRNA transfection.

References

[1] Ying, H; Zaks, TZ; Wang, RF; Irvine, KR; Kammula, US; Marincola, FM; Leitner, WW and Restifo, NP. Cancer therapy using a self-replicating RNA vaccine. *Nat. Med.*, 1999, 5, 823-27.

[2] Luo, D and Saltzman, WM. Synthetic DNA delivery systems. *Nat. Biotechnol.*, 2000, 18, 33-37.

[3] Sawai, K; Ohno, K; Iijima, Y; Levin, B and Meruelo, D. A novel method of cell-specific mRNA transfection. *Mol. Genet. Metab.*, 1998, 64, 44-51.

[4] Van Tendeloo, VF; Ponsaerts, P; Lardon, F; Nijs, G; Lenjou, M; Van Broeckhoven, C; Van Bockstaele, DR and Berneman, ZN. Highly efficient gene delivery by mRNA electroporation in human hematopoietic cells: superiority to lipofection and passive pulsing of mRNA and to electroporation of plasmid cDNA for tumor antigen loading of dendritic cells. *Blood*, 2001, 98, 49-56.

[5] Martinet, W; Schrijvers, DM and Kockx, MM. Nucleofection as an efficient nonviral transfection method for human monocytic cells. *Biotechnol. Lett.*, 2003, 25, 1025-29.

[6] Van de Parre, TJL; Martinet, W; Schrijvers, DM; Herman, AG and De Meyer, GRY. MRNA but not plasmid DNA is efficiently transfected in murine J774A.1 macrophages. *Biochem. Biophys. Res. Commun.*, 2005, 327, 356-60.

[7] Weir, JP and Meltzer, MS. Transfection of Human-Immunodeficiency-Virus Type-1 Proviral Dna Into Primary Human Monocytes. *Cell. Immunol.*, 1993, 148, 157-65.

[8] Shimokawa, T; Okumura, K and Ra, C. DNA induces apoptosis in electroporated human promonocytic cell line U937. *Biochem. Biophys. Res. Commun.*, 2000, 270, 94-99.

[9] Mendez, R and Richter, JD. Translational control by CPEB: A means to the end. *Nat. Rev. Mol. Cell Biol.*, 2001, 2, 521-29.

[10] Lenz, P; Bacot, SM; Frazier-Jessen, MR and Feldman, GM. Nucleoporation of dendritic cells: efficient gene transfer by electroporation into human monocyte-derived dendritic cells. *FEBS Lett.*, 2003, 538, 149-54.

[11] Trompeter, HI; Weinhold, S; Thiel, C; Wernet, P and Uhrberg, M. Rapid and highly efficient gene transfer into natural killer cells by nucleofection. *J. Immunol. Methods*, 2003, 274, 245-56.

[12] Ganesan, LP; Wei, G; Pengal, RA; Moldovan, L; Moldovan, N; Ostrowski, MC and Tridandapani, S. The Serine/Threonine Kinase Akt Promotes Fc gamma Receptor-Mediated Phagocytosis in Murine Macrophages through the Activation of p70S6 Kinase. *J. Biol. Chem.*, 2004.

[13] Mitchell, DA and Nair, SK. RNA-transfected dendritic cells in cancer immunotherapy. *J. Clin. Invest.*, 2000, 106, 1065-69.

[14] Ponsaerts, P; Van den, BG; Cools, N; Van Driessche, A; Nijs, G; Lenjou, M; Lardon, F; Van Broeckhoven, C; Van Bockstaele, DR; Berneman, ZN and Van Tendeloo, VF. Messenger RNA electroporation of human monocytes, followed by rapid in vitro differentiation, leads to highly stimulatory antigen-loaded mature dendritic cells. *J. Immunol.*, 2002, 169, 1669-75.

[15] Dokka, S; Toledo, D; Shi, XL; Ye, JP and Rojanasakul, Y. High-efficiency gene transfection of macrophages by lipoplexes. *Int. J. Pharm.*, 2000, 206, 97-104.

[16] Gao, X and Huang, L. Potentiation of cationic liposome-mediated gene delivery by polycations (vol 35, pg 1027, 1996). *Biochemistry*, 1996, 35, 9286.

[17] Sorgi, FL; Bhattacharya, S and Huang, L. Protamine sulfate enhances lipid-mediated gene transfer. *Gene Ther.*, 1997, 4, 961-68.

[18] Heiser, WC. Gene transfer into mammalian cells by particle bombardment. *Anal. Biochem.*, 1994, 217, 185-96.

[19] Arkusz, J; Stepnik, M; Trzaska, D; Dastych, J and Rydzynski, K. Assessment of usefulness of J774A.1 macrophages for the assay of IL-1 beta promoter activity. *Toxicol.* In Vitro, 2006, 20, 109-16.

[20] Escher, G; Hoang, A; Georges, S; Tchoua, U; El Osta, A; Krozowski, Z and Sviridov, D. Demethylation using the epigenetic modifier, 5-azacytidine, increases the efficiency of transient transfection of macrophages. *J. Lipid Res.*, 2005, 46, 356-65.

[21] Stepinski, J; Waddell, C; Stolarski, R; Darzynkiewicz, E and Rhoads, RE. Synthesis and properties of mRNAs containing the novel "anti-reverse" cap analogs 7-methyl(3 '-O-methyl)GpppG and 7-methyl(3 '-deoxy)GpppG. *RNA*, 2001, 7, 1486-95.

[22] Rupprecht, AP and Coleman, DL. Transfection of adherent murine peritoneal-macrophages with a reporter gene using Deae-Dextran. *J. Immunol. Methods,* 1991, 144, 157-63.

[23] Gibson, B; Duffy, AM; Fogerite, SG; Krause-Elsmore, S; Lu, RY; Shang, GF; Chen, ZW; Mannino, RJ; Bouchier-Hayes, DJ and Harmey, JH. A novel gene delivery system for mammalian cells. *Anticancer Res.*, 2004, 24, 483-88.

[24] Hashimoto, M; Morimoto, M; Saimoto, H; Shigemasa, Y; Yanagie, H; Eriguchi, M and Sato, T. Gene transfer by DNA/mannosylated chitosan complexes into mouse peritoneal macrophages. *Biotechnol. Lett.*, 2006, 28, 815-21.

In: New Messenger RNA Research Communications
Editor: Lee B. Kwang, pp. 191-201

ISBN 978-1-60021-488-2

Chapter XIII

Alternative Coding Potential of Mammalian mRNAs

Alex V. Kochetov[1,2]
[1]Institute of Cytology and Genetics, Lavrentieva 10, Novosibirsk 630090, Russia
[2]Novosibirsk State University, Novosibirsk 630090, Russia

Abstract

It is widely suggested that a eukaryotic mRNA typically contains one translation start site and encodes a single functional protein product. However, according to the linear scanning model, 40S ribosomal subunits can either initiate translation at the 5'-proximal AUG codon in a suboptimal context or miss it and initiate translation at downstream AUG(s), thereby producing several proteins. Earlier experiments suggested that some of these additional protein variants could be functionally important. I analyzed the potential alternative translation start sites in a sample of mouse cDNAs. It was found that 42% of the sample was characterized by AUG-containing 5'-untranslated regions and 14% were characterized by a suboptimal context of an annotated start codon. It is likely that many mouse mRNAs can additionally encode protein products either unrelated to annotated proteins or representing N-end truncated or N-end extended variants of the annotated protein. Statistical analysis of alternative ORFs features also demonstrated their potential functionality. In particular, there is a considerable difference in predicted subcellular localizations of many annotated mouse proteins and their potential N-end extended or N-end truncated variants. These results suggest that in mammalian cells additional AUG codons can frequently be used as alternative translation start sites to synthesize proteins possessing new functional properties. The correctness of translation start site(s) annotation, possible polypeptide heterogeneity, and the related mRNA features are discussed.

INTRODUCTION

Translation of most eukaryotic mRNAs is likely to be initiated by a linear scanning mechanism, although some other mechanisms are also possible (Kozak, 2005; Jackson, 2005). According to the scanning model, 40S ribosomal subunits can either initiate translation at the 5'-proximal AUG codon in a suboptimal context or miss it and initiate translation at downstream AUG (dAUG). It has been shown that recognition of the translation start codon by eukaryotic ribosomes depends on its nucleotide context. For mammalian mRNAs, the most important elements of the AUG context are purine at position -3 and guanine at position +4 (Kozak, 2005; Pisarev et al., 2006). It was also proposed that eukaryotic ribosomes could reinitiate translation at the downstream open reading frame (dORF). The efficiency of reinitiation depends on the size of upstream ORF (uORF) and the distance between uORF and dORF (Kozak, 2001; Poyry et al., 2004; Jackson, 2005).

It has been shown that eukaryotic mRNAs frequently have AUG-containing 5' untranslated regions (5'-UTRs) (Suzuki et al., 2000; Rogozin et al., 2001). It is suggested that upstream AUGs (uAUGs), uORFs, and the encoded peptides can often be used to control mRNA translation efficiency (Kochetov et al., 1998; Rogozin et al., 2001; Meijer and Thomas, 2002; Raney et al., 2002; Vattem and Wek, 2004; Kozak, 2005). Eukaryotic mRNAs are also often characterized by a suboptimal context of the translation start site (TSS) (Rogozin et al., 2001), which provides an opportunity for initiation of translation at closely located downstream AUG triplets. In particular, initiation of translation at several closely located in-frame AUGs could result in a synthesis of protein variants with different N-ends (N-end truncated or N-end extended) and these additional protein variants could be of functional importance (Tanaka and Ueda, 2000; Watanabe et al., 2001; Pedrajas et al., 2002; Logette et al., 2003; Oritani et al., 2003; Kochetov et al., 2003; 2005; Outten and Culotta, 2004; Kochetov and Sarai, 2004; Kaipio et al., 2005). Several small proteins encoded by 5'-UTR-located uORFs or CDS-located dORFs were detected in eukaryotic cells (Bab et al., 1999; Stallmeyer et al., 1999; Diba et al., 2001; Jousse et al., 2001; Oyama et al., 2004; Smith et al., 2005). However, functional roles of either uORF (dORF) nucleotide sequences or peptides/proteins potentially encoded in the vast majority of cases are unclear.

Thus, according to the current point of view on translation initiation mechanisms, eukaryotic mRNAs can contain alternative TSSs and encode more than one protein product. It is principally possible that translation can start at two or more closely located AUG codons. However, most eukaryotic mRNAs annotated in nucleotide sequence databanks are characterized by a single predicted TSS and a single protein encoded: alternative TSS(s) and corresponding ORF(s) are very rarely annotated and their protein products are commonly not considered as potential constituents of eukaryotic proteomes.

I extracted a sample of mouse cDNAs potentially containing alternative TSS(s), i.e. those with AUG-containing 5'-UTRs or suboptimal contexts of the annotated translational start codons. It was found that 42% of the sample were characterized by AUG-containing leader regions and 14% were characterized by a suboptimal context of annotated TSSs. It is likely that many mouse mRNAs can additionally encode protein products either unrelated to annotated proteins or representing N-end truncated or N-end extended variants of the annotated protein. Statistical analysis of alternative TSSs and ORFs features also demonstrated their potential functionality. In particular, there is a considerable difference in the predicted subcellular localizations of many annotated mouse proteins and their potential N-end extended and N-end truncated variants.

METHODS

Overall, 33 163 EMBL entries were obtained at http://srs.ebi.ac.uk/ using the following search fields and terms: "Organism", *Mus musculus*; "Molecule", mRNA; "FtKey", CDS (coding DNA sequence); and "Description", complete CDS. Of these, 27788 nucleotide sequences contained both the complete coding parts and 5'-UTRs shorter than 1000 and longer than 15 nucleotides (length limitation was used to avoid non-functional partially processed sequences (Kozak, 2005)). ReadSeq program (version 2.1.3; available at: http://iubio.bio.indiana.edu/soft/molbio/readseq/java/) was used for the samples preparation. Statistica for Windows package (Statsoft^{TM}, Tulsa, USA) was used for statistical analyses. We also developed a set of *ad hoc* programs for the selection of 5'-UTR-located ORFs, N-end extended and N-end truncated protein variants, etc. Subcellular localizations of proteins were evaluated by *TargetP* software (Emanuelsson et al., 2000) used with default parameters.

RESULTS

mRNA Characteristics

Sequence characteristics of 5'-UTR sample extracted from EMBL databank were analyzed (Table 1a). It was found that the general features (average length, base composition, and uAUG content) coincided well with earlier reports (Suzuki et al., 2000; Rogozin et al., 2001; Kochetov et al., 2005) (data not shown). Overall, 42% of the mouse cDNA 5'-UTRs contained AUG triplet(s). It is likely that some 5'-UTRs with a high AUG content may be translated by special mechanisms or correspond to non-functional poorly translated mRNAs (Kozak, 2005). In order to avoid them, I took into account a subsample of mRNAs containing only one uAUG. Two types of uORFs were selected for further analysis (Table 1a): (i) uORFs encoded N-end extensions to annotated proteins (i.e., where uAUG was located in frame with the annotated CDS and no termination signals were located between uAUG and the annotated TSS); (ii) other uORFs encoded unrelated peptides/small proteins.

Table 1a. Description of uORF sequence samples.

cDNAs with intact CDS	5'UTRs with different numbers of uAUGs, %			uORFs*	uORFs* overlapped with CDS,%	5'-end extensions,%**
	0	**1**	**2**			
27788	58	18	8	3047	51	12

*uORFs larger than 27 nucleotides were taken into account

**N-end extensions larger than 3 aa were taken into account

An mRNA sample containing an annotated translational start codon in a suboptimal context was used for isolation of dORFs started from the next downstream AUG codon. These dORFs were also classified into in-frame (encoded N-end truncated protein variant) and out-of-frame encoded unrelated peptides/small proteins (table 1b).

Table 1b. Description of dORF sequence samples.

cDNAs with suboptimal TSS, total	Downstream TSS features:	
	In-frame with CDS*	Out of CDS frame**
3955	1463	1376

*downstream AUG positioned at a distance greater than 3 codons were taken into account
**downstream ORFs larger than 27 bp were taken into account

Distribution of AUG Triplets at the CDS Beginning

The positional frequencies of AUG codons were calculated at the beginning of coding sequences in samples with optimal and suboptimal start codon contexts (figure 1). It was found that mouse mRNAs with suboptimal start codon contexts were characterized by a considerably higher frequency of in-frame AUGs within 5'-proximal CDS fragment spanning *ca.* 14 codons. The distribution of out-of-frame AUG triplets along this region was also analyzed.

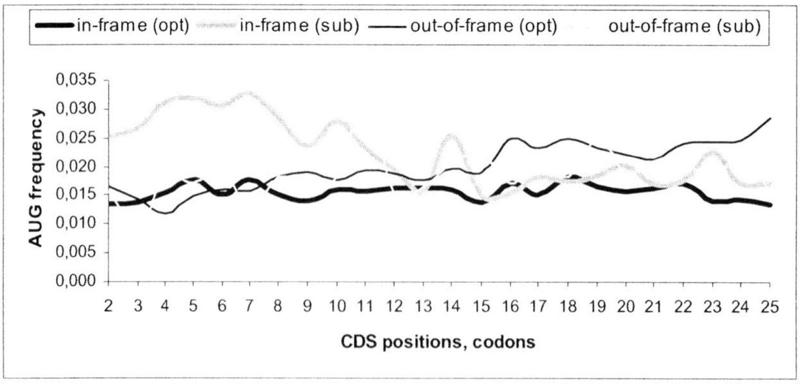

Figure 1. Average frequencies of AUG triplets at the beginning of the mouse CDSs (from 2nd to 25th codons) with either optimal (opt, purine at position –3) or suboptimal (sub, pyrimidine at position –3) contexts. dAUGs were located either in the CDS frame (i.e., they represented potential start sites for N-end extended protein variants) or in +1 frame (i.e., they represented potential start sites for unrelated peptides or small proteins). Frequencies of in-frame dAUGs located in mRNA samples with optimal and suboptimal contexts differ significantly in positions 2-12,14,15 ($P < 0.05$ according to *t*-test).

One can see that the distributions of average frequencies of in-frame and out-of-frame AUGs differ (figure 1): in-frame AUG codons are significantly more frequent at the CDS beginning if annotated TSSs are located in suboptimal contexts.

In general, the mRNAs with suboptimal TSSs were characterized by a significantly higher average frequency of closely located in-frame dAUGs than the mRNAs with optimal TSSs or the average AUG frequency expected in the assumption of a random codon choice (0.016)

uORF/dORF Features

An average size of uORF- and dORF-encoded peptides were 21 and 28 amino acids (respectively; 5'-end extended and 5'-end truncated CDS variants were excluded from these calculations (table 2)). uORFs possibly encoding N-end extended protein parts have an average length of N-extension equal to 17 amino acids which is quite enough to provide an N-end extended protein variant with some new features (*e.g.*, it may change a protein subcellular localization). Similarly, an average size of N-end fragment absent in N-end truncated protein variants encoded by in-frame dORFs was 19 amino acids (table 2).

Table 2. Average size of uORFs and dORFs started from alternative start codons in mouse mRNAs.

	Sample size	Average size (nucleotides; ± std. dev.)
uORFs*	4368	63 ± 78
5'-end extension**	393	51 ± 51
dORFs*	2281	84±93
5'-end truncations**	1674	57±62

*5'-end extensions were excluded from this calculation.

**5'-end extension, additional 5'-end portion of CDS located between in-frame uAUG and annotated translation start site; 5'-truncation, CDS portion located between annotated start site in a suboptimal context and the nearest (in-frame) dAUG

Subcellular Targeting of Annotated Proteins and Their Putative N-End Extended and N-End Truncated Variants

It was found that 12% of the uORFs were located in-frame with CDS and no termination signals occurred between uAUG and CDS. In total, 393 mouse mRNAs were selected for further analysis. To evaluate possible functional role of N-end extended proteins, I compared sorting of the annotated and N-end extended variants using the software *TargetP* (Emanuelsson et al., 2000), which allowed to predict the occurrence of either mitochondrial targeting peptide (MTP) or signal peptide (SP) (table 3). It is evident that the predicted sorting of the annotated and N-end extended proteins differs greatly: 119 out of 393 mouse mRNAs (30%) can encode additional protein variants targeted to different compartments in comparison with the annotated proteins.

Table 3. Subcellular localizations of annotated mouse proteins and their potential N-end extended variants (%) predicted with *TargetP* program (Emanuelsson et al., 2000).

Annotated		N-end extended		
Location	Size of fraction, %	MTP*	SP*	Others
MTP*	9	6	0.5	2.5
SP*	35	7	22.5	5.5
Other	56	11	4	41
Total	100	24	27	49

*MTP, mitochondria targeting peptide; SP, secretory peptide.

In total, 51% of N-end extended and 44% of annotated proteins were predicted to contain MTP or SP, 70% of N-end extended and annotated proteins were characterized by the same predicted targeting, and 7% of N-end extended and annotated counterparts were characterized by changed predicted sorting (from MTP-containing to SP-containing protein form or *vice versa*).

Similarly, 3955 out of 27616 mRNAs contain the annotated start codon in a suboptimal context (i.e., C or U are located at position –3 upstream of AUG). In total, 1463 putative N-end truncated proteins (i.e., where dAUGs were located in the CDS frame) were collected. Predicted subcellular localizations of the annotated and N-end truncated protein variants were compared with the aid of *TargetP* program (Table 4). It was found that ca. 26% of the annotated and N-end truncated protein variants were targeted to distinct subcellular compartments (table 4).

Table 4. Subcellular localization of annotated mouse proteins and their putative N-end truncated variants (%) predicted with the *TargetP* program (Emanuelsson et al., 2000).

Annotated		N-end truncated		
Location	Size of fraction,%	MTP*	SP*	Others
MTP*	16	5	2	9
SP*	22	1	15	6
Other	62	4	4	54
Total	100	10	21	69

*MTP, mitochondria targeting peptide; SP, secretory peptide.

In total, 15% of N-end truncated proteins miss targets of their annotated variants (*i.e.*, *TargetP* prediction changed from MTP- or SP-containing to "Others"), 20% of N-end truncated proteins retained localization of their full counterparts (MTP or SP), 8.5% of N-end truncated proteins acquired sorting signals *de novo* (*i.e.*, *TargetP* prediction was changed from "Other" to MTP- or SP-containing) and 3% changed their predicted sorting (from MTP-containing to SP-containing protein form or *vice versa*).

In addition, 411 mouse uORF-encoded and 345 dORF-encoded peptides are larger than 50 amino acids, that is these peptides can be considered as small proteins.

CONCLUSION

It is known that a large part of eukaryotic mRNAs is characterized by either a suboptimal context of the translation start site or the presence of an AUG triplet within 5' UTR (Suzuki et al., 2000; Rogozin et al., 2001; Kochetov et al., 2005; Tables 1a and 1b in this chapter). It is likely that at least some mRNAs may produce two or more functional peptides and proteins due to a leaky scanning or reinitiation (Bab et al., 1999; Stallmeyer et al., 1999; Diba et al., 2001; Jousse et al., 2001; Oyama et al. 2004; Wang and Rothnagel, 2004; Smith et al., 2005; Kozak, 2005). However, the actual role of this mechanism was not systemically studied, and the AUG codons at the beginning of CDS were not commonly taken into account in the evaluation of mRNA translation efficiency and coding potential.

I used in analysis the sample of mouse mRNAs containing single uAUG (to avoid the partially processed transcripts and mRNAs with a complex control of translation initiation process) and the sample of mRNAs characterized by a suboptimal context of the annotated translation start site. Several variants of alternative ORFs are possible in these cases (figure 2): either encoding peptide or small protein (uORF1, uORF2, dORF1), or encoding N-end extended (uORF3) or N-end truncated (dORF2) versions of a protein annotated in the nucleotide sequence databank..

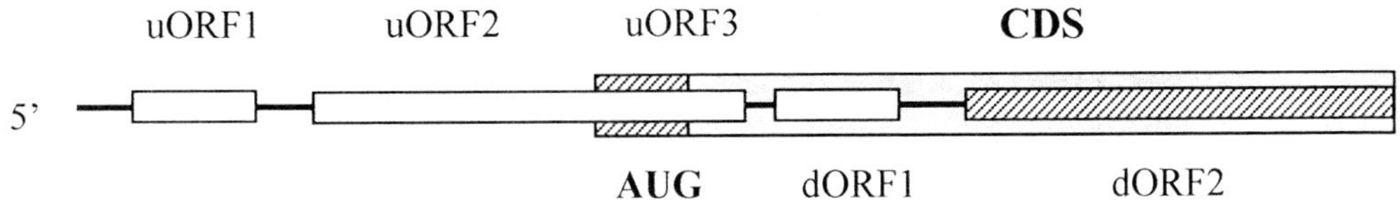

Figure2. Alternative translation start sites and open reading frames. uORFs start from AUG codons located within 5'-untranslated region and can be either 5'-UTR-restricted (uORF1; white box), partially overlap with CDS (uORF2; white box) or fuse with CDS (*i.e.*, if uAUG lies in the CDS frame and no termination codons are located between this uAUG and the annotated start codon; uORF3, striped box). dORFs start from AUG codons located downstream the annotated start codon in a suboptimal context and can either encode small protein (dAUG is located out of the CDS frame: dORF1, white box), or encode N-end truncated protein variant (dAUG is located in frame with CDS: dORF2, striped box). Annotated CDS is marked by grey color.

It may be assumed that closely located in-frame AUG codons can be used as additional start sites to recruit 40S ribosomal subunits missing the proximal start codon in a suboptimal context (Kozak, 2005). The usage of a closely located downstream in-frame AUG codon as an alternative TSS could result in the additional synthesis of a slightly truncated isofunctional protein variant: this could increase the rate of protein synthesis and prevent the potentially harmful initiation of translation at out-of-frame AUGs (Kochetov et al., 2003, 2005; Kochetov, 2005; Kozak, 2005). To test this assumption, the samples of mouse mRNAs with optimal and suboptimal contexts of the annotated start codons were compared. If the downstream AUG codons are utilized as alternative translation start sites, their occurrence has to correlate with the start codon context: suboptimal context should be accompanied by a higher frequency of alternative TSSs. It was found that the contextual organization of CDS beginning (*ca.* 15 codons) in mRNAs with a suboptimal start codon context was characterized by a specific increase in the frequency of in-frame AUG codons (Fig. 1). Note that the average frequency of out-of-frame AUG triplets in mouse mRNAs was lower within this 5'-terminal CDS region and did not depend on the start codon context (Fig. 1). In my opinion, it is unlikely that a correlation between the context of annotated translation start codon and the frequency of downstream in-frame AUG codons can be associated with the reasons other than the use of downstream in-frame AUGs as alternative translation start sites. This observation coincides well with the hypothesis about functional significance of downstream AUG codons: it is likely that such a mechanism is used to increase translation initiation of some ORFs with a suboptimal context of the 5'-proximal start AUG codon, and some eukaryotic proteins are heterogeneous at their N-ends (Kochetov, 2005; Fig.1 in this chapter). For example, 787 of 3045 mouse mRNAs with suboptimal contexts of annotated start codons contained at least one in-frame AUG codon within the nine downstream positions and most of these dAUGs were located in the optimal context (mRNAs with more than two uAUGs were eliminated from this calculation).

N-end extended and N-end truncated variants of annotated proteins can also possess new functional properties (Kochetov and Sarai, 2004; Kochetov et al., 2005). It was experimentally demonstrated that an additional translation start site could be used to target some proteins to several locations (e.g., Tanaka and Ueda, 2000; Watanabe et al., 2001; Pedrajas et al., 2002; Logette et al., 2003; Leissring et al., 2003; Outten and Culotta, 2004; Kaipio et al., 2005). To evaluate the functional significance of putative alternative protein variants, the predicted sorting of 1856 annotated proteins and their N-end extended or N-end truncated counterparts was compared. It was found that 499 of 1856 mRNAs analyzed were characterized by different predicted subcellular localizations for annotated protein variants and their N-end extended and N-end truncated counetrparts. It seems likely that at least 2% of mouse genes (considering that the mouse genome contains ca. 25,000 genes) might encode additional protein variants targeted to other subcellular compartments. Note that this evaluation is certainly understated: I did not take into account the huge number of uORFs located within 6700 mouse mRNAs with more than one uAUG. I also did not take into consideration the nucleotide context of dAUGs: it is quite possible that the ribosomes can initiate translation further downstream if the proximal dAUG is located in a suboptimal context. N-end extended and N-end truncated protein variants may differ from their annotated counterparts in the characteristics other than the subcellular targeting (e.g., Bab et al., 1999; Oritani et al., 2003; Smith et al., 2005). Thus, such a mechanism could have an important impact on the mouse cellular and organelle proteomes.

In the context of this assumption, some mRNAs can be characterized by complex translation initiation signals containing two or more AUG codons and their translational efficiencies and coding potentials should be re-evaluated. Recent investigation showed that many eukaryotic genes yielded transcript(s) that translate into several and often very numerous families of polypeptide species (Kettman et al., 2002). Initiation of translation at alternative AUG codons could also result in a synthesis of additional protein variants possessing new functional properties. Study of the mRNA features correlating with the TSS "strength" (including those described here) could be useful for evaluation of translation initiation rate and new functional forms of eukaryotic proteins.

ACKNOWLEDGEMENT

I thank Nikolay Kolchanov, Akinori Sarai, Lev Kisselev and Igor Rogozin for helpful discussions. This work was supported by the Russian Foundation for Basic Research (grant No. 05-04-48207) and the Programs of Russian Academy of Sciences (Origin and Evolution of Biosphere and Dynamics of Gene Pools). I thank SD RAS Complex Integration Program (No. 5.3), and Ministry of Science and Education of the Russian Federation (grant No. 528.2006.4) for partial support.

REFERENCES

Bab, I; Smith, E; Gavish, H; Attar-Namdar, M; Chorev, M; Chen, YC; Muhlrad, A; Birnbaum, MJ; Stein, G; Frenkel, B. Biosynthesis of osteogenic growth peptide via alternative translational initiation at AUG85 of histone H4 mRNA. *J Biol Chem*, 1999 274, 14474-14481.

Diba, F; Watson, CS; Gametchu, B. 5'UTR sequences of the glucocorticoid receptor 1A transcript encode a peptide associated with translational regulation of the glucocorticoid receptor. *J Cell Biochem*, 2001 81, 149-161.

Emanuelsson, O; Nielsen, H; Brunak, S; von Heijne, G. Predicting subcellular localization of proteins based on their N-terminal amino acid sequence. *J Mol Biol*, 2000 300, 1005-1016.

Jackson, RJ. Alternative mechanisms of initiating translation of mammalian mRNAs. *Biochem Soc Trans*, 2005 33, 1231-1241.

Jousse, C; Bruhat, A; Carraro, V; Urano, F; Ferrara, M; Ron, D; Fafournoux, P. Inhibition of CHOP translation by a peptide encoded by an open reading frame localized in the chop 5'UTR. *Nucleic Acids Res*, 2001 29, 4341-4351.

Kaipio, K; Kallio, J; Pesonen, U. Mitochondrial targeting signal in human neuropeptide Y gene. *Biochem Biophys Res Commun*, 2005 337, 633-640.

Kettman, JR; Coleclough, C; Frey, JR; Lefkovits, I. Clonal proteomics: one gene - family of proteins. *Proteomics*, 2002 2, 624-631.

Kochetov, AV; Ischenko, IV; Vorobiev, DG; Kel, AE; Babenko, VN; Kisselev, LL; Kolchanov, NA. Eukaryotic mRNAs encoding abundant and scarce proteins are statistically dissimilar in many structural features. *FEBS Lett*, 1998 440, 351-355.

Kochetov, AV; Kolchanov, NA; Sarai, A. Interrelations between the efficiency of translation start sites and other sequence features of yeast mRNAs. *Mol Genet Genomics* 2003 270, 442-447.

Kochetov, AV; Sarai, A. Translational polymorphism as a potential source of plant proteins variety in *Arabidopsis thaliana*. *Bioinformatics*, 2004 20, 445-447.

Kochetov, AV. AUG codons at the beginning of protein coding sequences are frequent in eukaryotic mRNAs with a suboptimal start codon context. *Bioinformatics*, 2005 21, 837-840.

Kochetov, AV; Sarai, A; Rogozin, IB; Shumny, VK; Kolchanov, NA. The role of alternative translation start sites in generation of human protein diversity. *Mol Genet Genomics,* 2005 273, 491-496.

Kozak, M. Regulation of translation via mRNA structure in prokaryotes and eukaryotes. *Gene*, 2005 361, 13-37.

Kozak, M. Constrains on reinitiation of translation in mammals. *Nucleic Acids Res*, 2001 29, 5226-5232.

Leissring, MA; Farris, W; Wu, X; Christodoulou, DC; Haigis, MC; Guarente, L; Selkoe, DJ. Alternative translation initiation generates a novel isoform of insulin-degrading enzyme targeted to mitochondria. *Biochem J,* 2001 383, 439-446.

Logette, E; Wotawa, A; Solier, S; Desoche, L; Solary, E; Corcos, L. The human caspase-2 gene: alternative promoters, pre-mRNA splicing and AUG usage direct isoform-specific expression. *Oncogene,* 2003 22, 935-946.

Meijer, HA; Thomas, AAM. Control of eukaryotic protein synthesis by upstream open reading frames in the 5'-untranslated region of an mRNA. *Biochem J,* 2002 367, 1-11.

Oritani, K; Hirota, S; Nakagawa, T; Takahashi, I; Kawamoto, S; Yamada, M; Ishida, N; Kadoya, T; Tomiyama, Y; Kincade, PW; Matsuzawa, Y. T lymphocytes constitutively produce an interferonlike cytokine limitin characterized as a heat- and acid-stable and heparin-binding glycoprotein. *Blood,* 2003 101, 178-185.

Outten, CE; Culotta, VC. Alternative start sites in the *Saccharomyces cerevisiae* GLR1 gene are responsible for mitochondrial and cytosolic isoforms of glutathione reductase. *J Biol Chem,* 2004 279, 7785-7791.

Oyama, M; Itagaki, C; Hata, H; Suzuki, Y; Izumi, T; Natsume, T; Isobe, T; Sugano, S. Analysis of small human proteins reveals the translation of upstream open reading frames of mRNAs. *Genome Res*, 2004 14, 2048-2052.

Pedrajas, JR; Porras, P; Martinez-Galisteo, E; Padilla, CA; Miranda-Vizuete, A; Barcena, JA. Two isoforms of *Saccharomyces cerevisiae* glutaredoxin 2 are expressed in vivo and localize to different subcellular compartments. *Biochem J,* 2002 364, 617-623.

Pisarev, AV; Kolupaeva, VG; Pisareva, VP; Merrick, WC; Hellen, CU; Pestova, TV. Specific functional interactions of nucleotides at key -3 and +4 positions flanking the initiation codon with components of the mammalian 48S translation initiation complex. *Genes Dev,* 2006 20, 624-636.

Poyry; TA; Kaminski, A; Jackson, RJ. What determines whether mammalian ribosomes resume scanning after translation of a short upstream open reading frame? *Genes Dev,* 2004 18, 62-75.

Raney, A; Law, GL; Mize, GJ; Morris, DR. Regulated translation termination at the upstream open reading frame in S-adenosylmethionine decarboxylase mRNA. *J Biol Chem*, 2002 277, 5988-5994.

Rogozin, IB; Kochetov, AV; Kondrashov, FA; Koonin, EV; Milanezi, L. Presence of ATG triplets in 5' untranslated regions of eukaryotic cDNAs correlates with a "weak" context of the start codon. *Bioinformatics,* 2001 17, 890-900.

Smith, E; Meyerrose, TE; Kohler, T; Namdar-Attar, M; Bab, N; Lahat, O; Noh, T; Li, J; Karaman, MW; Hacia, JG; Chen, TT; Nolta, JA; Muller, R; Bab, I; Frenkel, B. Leaky ribosomal scanning in mammalian genomes: significance of histone H4 alternative translation *in vivo*. *Nucleic Acids Res*, 2005 33, 1298-1308.

Stallmeyer, B; Drugeon, G; Reiss, J; Haenni, AL; Mendel, RR. Human molybdopterin synthase gene: identification of a bicistronic transcript with overlapping reading frames. *Am J Hum Genet*, 1999 64, 698-705.

Suzuki, Y; Ishihara, D; Sasaki, M; Makagawa, H; Hata, H; Tsunoda, T; Watanabe, M; Komatsu, T; Ota, T; Isogai, T; Suyama, A; Sugano, S. Statistical analysis of the 5'untranslated region of human mRNA using "oligo-capped" cDNA libraries. *Genomics*, 2000 64, 286-297.

Tanaka, A; Ueda, M. Sorting of peroxisomal and mitochondrial carnitine acetyltransferase isozymes in the diploid yeast, *Candida tropicalis*. Cell Biochem Biophys, 2000 32, 139-146.

Vattem, AM; Wek, RC. Reinitiation involving upstream ORFs regulates ATF4 mRNA translation in mammalian cells. *Proc Natl Acad Sci USA*, 2004 101, 11269-11274.

Wang, X-Q; Rothnagel, JA. 5'-untranslated regions with multiple upstream AUG codons can support low-level translation via leaky scanning and reinitiation. *Nucleic Acids Res*, 2004 32, 1382-1391.

Watanabe, N; Che, F-S; Iwano, M; Takayama, S; Yoshida, S; Isogai, A. Dual targeting of spinach protoporphyrinogen oxidase II to mitochondria and chloroplasts by alternative use of two in-frame initiation codons. *J Biol Chem*, 2001 276, 20474-20481.

Index

A

D

G

H

I

J

K

L

M

N

O

S

T